THE GREEN SELF-BUILD BOOK

THE GREEN
SELF-BUILD BOOK

How to design and build your own eco-home

Jon Broome

GREEN BOOKS

First published in 2007
by Green Books Ltd
Foxhole, Dartington,
Totnes, Devon TQ9 6EB

Design by Rick Lawrence, samskara@onetel.com

Photographs are by the author as well as the following, who kindly made images available:

Architype, Nigel Corrie, Nick Grant of Elemental Solutions, John Brooke, Ralph Carpenter of Modece
Architects, Kevin McCabe, Brian Richardson, David Matzdorf, Hockerton Housing Project,
Karin Stockerl, Construction Resources, John Little of the Grass Roof Company

Text printed on Emerald FSC paper (75% recycled, 25% Forest Stewardship Council approved)

Printed and bound by Butler & Tanner, Frome, Somerset

British Library Cataloguing in Publication Data available on request

ISBN 978 1 903998 73 1

Contents

Acknowledgements

My wife Rona Nicholson for helpful suggestions on the text.

Those friends and relatives who helped us to build Shaws Cottages.

Thanks also to those who contributed to the other case studies, and allowed me to interview them and photograph their homes – in particular:

Nick Martin and Nick White at Hockerton, for their practical knowledge and experience of renewable energy systems, and for photographs of the scheme,

The Islington Self-Build Group for their perseverance in the face of unreasonable delays in securing funding,

John Little and Fiona Crumay, who built a stunningly economical home in Basildon,

Nick Grant and Sheila Herring for their hospitality, details of the construction and photographs,

Carol Coombes and John Brooke for construction photographs,

Ralph Carpenter of Modece Architects for a tour of his home in Suffolk and other hemp buildings in the area and for construction photographs,

and Kevin McCabe for taking time from his cob-building enterprise.

Brian Richardson was co-author with me of *The Self-Build Book*, and I am indebted to him for his excellent account of the use of The Pattern Language, which is reprinted here together with a new piece on a 'Place of Your Own'. Brian's delightful drawings illustrate this description, together with some of his photographs.

Jane Gosnell is co-author with me of the Sustainability Works website. Much of the sections of *The Green Self-Build Book* on energy, environmental impacts, health, waste and water are based on material which was first researched and written for the website, and I am indebted to Jane for her work on this.

A number of experts commented on the material in the Sustainability Works website, and I am indebted for their observations. The architect Pat Borer commented on the material on buildings; Energy Consultant John Willoughby commented on energy, and Nick Grant of Elemental Solutions contributed to the section on water and sewerage.

Many thanks to Ruth Ashby for preparing most of the construction drawings in the book.

Thanks for drawings of the Ealing steel-framed self-build house go to the designers Burd Haward Marston Architects.

Dedication

Dedicated to the memory of Colin Boyne:
architectural journalist, Trustee of the
Walter Segal Self-Build Trust,
and enthusiast for self-build.

Why self-build, and why build green?

Building your own home can be a very satisfying thing to do, and many people have gained a great sense of achievement from doing it. At the same time you can ensure that your home does not make unsustainable impacts on the environment. Homes currently consume around 30% of all the energy used in the UK, and this figure is rising. However, a home which does not rely on fossil fuel is perfectly feasible using current technology. At the same time, more and more people are becoming aware of the limitations of the mass housing market and are looking for homes which are well designed, better equipped, and more energy-efficient than the market provides.

What this book contains

The book describes examples of people who have successfully carried out green self-build projects, and shows what they have achieved. The examples include different forms of construction appropriate for self-builders: timber construction for speed and adaptability, steel for lightness and strength, straw for low embodied energy, earth-sheltered for storing energy, and cob for low cost. The book goes on to outline:

- the issues that you should consider when you design a house, including the need for a house to be capable of adapting to changing needs and expectation in the future

- sustainable methods of construction which are preferred for the walls, roof and floors and other principal elements of a house, together with the issues that have to be addressed to reduce the impact on the environment of building and occupying houses

- sources of materials and components which are suitable for low-environmental-impact construction

- the policy implications of a wider commitment to a more sustainable development process

What building your own home has to offer

My main aim is to inspire you to build for yourself. Organizing the designing and building of your own house is within the reach of us all. It is enjoyable and can have great economic, practical and social benefits. You can feel the satisfaction of making something really useful, and experience the excitement of dreaming about what your house will be like, how it will be laid out, what you are going to put in it and what it will be like to live in. You will see it slowly taking shape and will imagine the next steps in your mind and have a vision of the finished house. A handmade house is a pleasure to live in, and you will know every corner intimately. You will be able to afford a bigger, better house, arranged to suit your particular needs and desires, and you will be able to reduce your housing costs.

More and more people are becoming aware of the limitations of the mass housing market. It often offers expensive but poorly designed and equipped houses, badly built on characterless estates which are often located in areas which are not very desirable to live in. In many continental countries there is more awareness of how things could be otherwise – such as in Germany, where the housing market is not controlled to the same extent by speculative developers building for profit. The self-help sector there is four times larger than in Britain. Many people are able to commission a one-off house built in a village by the local builder to their particular requirements. Houses are larger, with basement storage and workshop space, with better kitchens and bathrooms, and are much more energy-efficient, with triple-glazing as standard.

This rising awareness of an alternative, together with greater prosperity, have made the self-help sector in Britain rise from a meagre 4,000 dwellings, or 2% of the private housing market, in 1980 – the lowest proportion of any developed economy anywhere, lower than in North America, Australia, Japan as well as the rest of Europe – to a level estimated to be

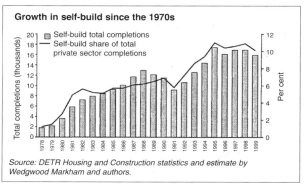

Growth in self-build since the 1970s

Source: DETR Housing and Construction statistics and estimate by Wedgwood Markham and authors.

18,000 dwellings or 11% of the market in 1999, and 25,000 or 15% at the time of writing. And it is this growing interest that has fuelled the success of television shows such as 'Grand Designs'.

I should make clear at this point what I mean by self-build. I am referring to those homes which are commissioned by the people who live in them. They may or may not carry out the construction work themselves. They may commission a builder to do the building work, or they may employ subcontractors directly to do all or part of the building work whilst they carry out the project management work that is normally carried out by the builder. What is significant is that if you organize but do not necessarily actually build yourself, you are nevertheless in control of the process; you decide how your home is designed and how much money you want to spend. You will save money if you do all or most of the work yourself, but it will involve a greater time commitment – which may have costs associated with it, through loss of earnings arising from part-time working, for example.

All sorts of different people self-build for all sorts of different reasons. Some people have come together in groups because they wish to create a community and live with other like-minded people; disabled people have self-built, relying on their able-bodied friends and relations to carry out most of the physical work, so that they could lay out the house to suit their particular needs and to have the adaptability to change the arrangements if necessary in the future.

I have worked with self-builders from all walks of life, and have shared their excitement and enjoyment of their individually designed houses. They have been from very different backgrounds and with very different personalities. They included young people, older people building for their retirement, single mothers, unemployed people, people with low incomes, and almost all of them have had no experience of building. They have enjoyed the satisfaction of making a home for their household which is designed to their needs and desires, well built, energy-efficient and inexpensive to live in, all

Top: Building together in Brighton.
Centre: Family self-build.
Bottom: A member of a black self-build housing co-operative.

achieved at less cost. There is no doubt in my mind that there are many people who, like them, would jump at the chance to organize building houses for themselves if they thought that it was a real possibility. Anyone can do it if they have the opportunities and determination – and are prepared to work hard.

Building your own home requires a great deal of effort and commitment, depending on how much of the building work you do yourself. This can be particularly difficult if you are doing it in your spare time and holding down a job. It can take over your life for a couple of years. It will put your relationships with your partner, children and friends under pressure. However, most people who have done it would tell you that the effort had been well worthwhile.

The benefits of a green approach to building

From the 1960s onwards there has been an impetus to reduce the use of energy because of concern that primary resources, and particularly oil, were going to run out. Since then the emphasis has changed to a concern about the level of pollution and the resulting global warming effect. Meanwhile, the Rio Earth Summit in 1992 broadened the areas of concern to include the ecology of the planet and the environment as a whole: the holes in the ozone layer, shortage of adequate water supplies, exploitation of the rainforests, reduction in biodiversity, and threats to human health posed by urban living. These are all now part of a global awareness of environmental issues which goes beyond the important issue of energy to include less tangible and more philosophical issues, for example those connected with health and stress. The ideas of sustainability and sustainable development have been developed as a framework for bringing energy, other finite resources and non-human species into balance with humanity – a humanity placed squarely within the environment, and not in any way separate from nature. This is the background against which the practice of sustainable design and building has developed.

Society's current way of life is not sustainable. It is estimated that if the population of the Earth consumed resources at the rate presently current in Britain, three Planet Earths would be required to supply them and to dispose of the waste being generated. Building can have an important role to play in addressing this state of affairs. It is estimated that the environmental impacts of buildings can be halved using current technology. The potential benefits of a sustainable approach to building include:

- Reducing the use of natural resources: fuel for heating and for generating electricity for lighting and power; materials and energy required to manufacture and transport the material used in the construction itself; and water for drinking, bathing and disposing of waste.

- Conserving man-made resources: ensuring that buildings have a long useful life, are built to last and to adapt to changing needs and expectations.

- Building social capital: people's capacity to organize and work together by making sure that they have an active part to play in the system.

The elements of green building

The government published the Stern Review of the Economics of Climate Change in November 2006, which concluded amongst other things that carbon emissions need to be reduced by at least 60% and possible as much as 80% by 2050 in order to stabilize the environment of the planet. The government published the Code for Sustainable Homes in December 2006, which set voluntary standards for reducing the environmental impacts of homes. The highest level of the code, Level 6, calls for the net carbon emissions from a home to be zero, and a target has been set for all new homes to achieve this standard by 2016. Some authorities, including the Welsh government and the Greater London Authority, have set more ambitious targets to achieve zero-carbon new homes by 2012.

Homes currently consume around 30% of all the energy used in the UK, and the figure is rising. However, a home which does not rely on imported energy is perfectly feasible using current technology. A substantial development of 82 homes called BedZed was completed in 2002 in South London as a demonstration of zero-emissions homes, and there are other examples shown in some detail in this book.

The energy used in heating, lighting and supplying power to a dwelling needs to be reduced, but the energy that is required to build it has to be addressed. This is the so-called embodied energy, the energy required to dig the raw materials out of the ground, transport them to a factory where they are turned into metal or plastic for example, and are then transported to another factory where they are manufactured into products and components such as windows, which in turn are transported to the site and incorporated into the building using power tools. This embodied energy can account for up to 50% of the total energy consumed during the life of a low-energy-use home. It can be reduced substantially by using natural materials such as timber where the energy from the sun does the work, materials from nearby which do not require to be transported far before they can be used (e.g. timber from the UK rather than from Siberia), by reusing materials such as second-hand flooring, and by using products made from recycled materials such as cellulose fibre insulation from newspaper.

Left: Based on a Danish model, the first co-housing development in Britain, in Stroud in Gloucestershire, consists of 35 private houses and flats with a shared common house at the heart. Right: BedZed, a pioneering eco-development in south London.

It is estimated that around 50% of the resources taken from nature are building-related. More efficient ways of building and using renewable resources such as timber can substantially reduce the resources used in building.

The demand for water for buildings, industry and agriculture doubled over the 20 years from 1970 to 1990, resulting in water shortages in some parts of Britain over the last few years. The effect of climate change is uncertain, but parts of the country are likely to get less rainfall. Domestic consumption accounts for around 65% of the total, and it has been demonstrated that water-efficiency measures can reduce domestic consumption by up to 50%, so water conservation can have a significant effect on reducing consumption of this resource.

It is also estimated that 50% of waste is generated in the building sector. Again, reusing materials and recycling waste can reduce the amount of waste generated significantly.

Modern construction uses many materials and products that are potentially harmful to health: adhesives, mastics, additives in plastics and gas emissions from timber-board products are all potentially harmful when people are exposed to them over long periods. The use of all these products can be avoided.

Sustainable buildings are designed to be durable and flexible so that they have a long useful life. The occupants of homes in sustainable communities should have an active role in the management of their homes to ensure that they are cared for and maintained into the future.

Sustainable construction strikes a balance between the potentially conflicting demands of the use of energy, other resources and ecology. Glass, for example, requires a significant amount of energy to produce by melting sand, and yet it can be used to trap the energy of the sun and put it to direct use. An urban wildlife garden supports far more species than the rural desert created by the use of pesticides and fertilizers. Sustainable development requires the application of knowledge in the pursuit of a healthier and more fulfilling way of living which can be sustained into the future.

Sustainable homes will:

- Provide greater comfort

- Reduce running costs

- Reduce the need for maintenance

- Create less pollution and use less of our scarce resources – fuel, materials and water

- Create less waste

- Provide healthy living environments

- Last longer and adapt to changing needs and expectations

- Support sustainable communities.

The environmental crisis – and the role of construction and housing in reducing its impact – is an issue that will not go away. International policy is being developed which is informing national, regional and local strategies. Time is short. Even if we were to prevent further emissions into the atmosphere from now on, average temperatures will continue to rise until the end of the century. It is imperative that action is taken now. Remember that a new building built today will still be consuming energy well into the future – and so it had better be a zero-emissions building.

Looking beyond the issue of sustainable building, let us not forget that to bring humanity into balance with Planet Earth we must consider the resources used to bring food across the world, such as to satisfy our desire to eat strawberries out of season; how we can holiday without jetting across the world; and how we can moderate our use of the car, for example. Building in a sustainable way has an important part to play, but it has to be seen in the context of our whole lifestyles.

Above: A member of a group of disabled self-builders who organized a development in Colchester. They relied on their able-bodied friends and relatives to do most of the building work, but they were able to design houses to suit their particular needs.
Below: One of the houses built by disabled self-builders in Colchester.

Who has successfully built a green self-build house?

This chapter aims to capture a sense of what it is like to build your own home: to illustrate the kind of issues that you will have to deal with, and to set the experience of building for yourself within the context of your day-to-day life. It also illustrates the wide range of reasons why people build for themselves: to get an affordable home, to reduce the environmental impacts of their lifestyle as far as possible, and to demonstrate to others what can be achieved. The choice of examples also illustrates the range of techniques of construction which have been developed to reduce the environmental impact of building.

2.1 My house built using timber poles

Above: Jon built with his sister a house designed by Walter Segal.

This case study describes building my own house in South London. It is a story which I am very familiar with, and is described in more detail than the examples that follow. It conveys a good sense of the issues faced in a green self-build project, and a sense of what it is like to build your own home.

I had worked with the pioneering self-build architect Walter Segal on the original Lewisham self-build houses, and had gone on to build one of them with my sister. She moved to Yorkshire and got married, and meanwhile I had a relationship with Rona Nicholson, who moved into the house in Segal Close a week or two before our first child was born. It was to be hoped that Alex would not be the only one. We got to thinking that we would be wanting more space both inside and outside the house one day soon. Segal Close enjoys a magnificent view over other people's enormously long gardens. Although this gives the impression of great spaciousness, more like living in deep country than in the city, the house itself only has a very small garden. We approached

the neighbours with a view to buying the end of their garden to put the extendibility of a Segal house into practice, but they were not willing to negotiate. There followed a dispiriting and increasingly half-hearted search for a site to build a new house from scratch. We looked at overpriced scraps of land around South London, and were about to abandon the idea when we learned of a garden at the back of a friend's house half a mile away in Forest Hill, which its owner was keen to sell. This is how we started our new self-build project.

I am going to recount the intervening three years as a self-builder's diary, and thereby hope to capture some of the feeling of what it is like to build a house. There is the strange blend of dreams and mundane work, how the process affects the different people involved, and I hope to bring to life that mysterious process by which idea is turned into reality through the exercise of skill and perseverance.

The dream

I had built, but was not the designer of, Segal Close. I had designed for other self-builders but had not had to carry out the designs myself. I wanted to experience the intimate relationship between house and occupant that could come from fulfilling my dream of designing and building my own home. Rona had a slightly different vision which nevertheless amounted to much the same thing, which was to live in OUR house rather than MY house, a house that we had created together. Unlike the houses in Segal Close, it was to have no timber battens on the walls inside and be set in a large garden.

May 1991: the site

The site was a square of 30x30 metres, set at one end of a block of back gardens between houses adjacent to an alley. Half the area was occupied by a large lawn surrounded by mature trees and borders planted with roses, one quarter was an orchard with a pond, and the final quarter was an overgrown vegetable garden complete with greenhouse and shed. The area between the surrounding houses had been occupied with tennis courts during the twenties and thirties, and the shed was the old pavilion.

The site is a 15-minute walk from the station, with a train every 10 minutes to London Bridge, which is a 15-minute journey away. There is a bus, and some shops at the bottom of the road. I was working at London Bridge and Rona worked at home, so we didn't need to use the car very much. The shops included all the essentials: baker, greengrocer, butcher, newsagent, off-licence, post office and fish and chip shop. They have now been replaced by a hairdresser, nail parlour, interior design studio and antique shop, since a hypermarket opened up not far away. The park and the school are a short walk away, and there is a pub 30 yards away. I was much looking forward to the idea of living next to a pub, but it changed hands and I now walk to a pub nearby with live jazz in the bar. The site offered space for a large house set in a mature garden in an inner suburban location, with good access to the city and fairly good local facilities.

Rona fell in love with the site as I had done. It was an opportunity too good to miss, and so we resolved to proceed.

June: the offer

We sought advice on the value of the plot. This is related to size, position and outlook but most importantly to the development potential of the land. Enquiries at the Town Hall suggested that the Local Authority would only give planning permission for a single house on this parcel of land, but it could be a large one. We wanted three children's bedrooms, a bedroom for us, a spare bedroom, a family room, a more formal living room and a study – a total of eight habitable rooms. The cost of land in a particular area can be expressed as the cost of land per habitable room. At the time the range in London was between £5,000 and £7,000 per habitable room, depending on location. We offered £48,000 for the site, which was £6,000 per habitable room for the house we wanted to build. It was a difficult judgement, wanting to offer enough to secure the land but not more than necessary. This was all made possible because when Rona moved into Segal Close she sold a house, the proceeds of which allowed us to offer cash for the site. Some back-of-envelope sums suggested that our savings would go a good way towards paying for the building, and so we decided to proceed. To our great relief, the offer was accepted.

July: the solicitors

A number of critical questions had to be answered before we would know if it was possible to build the house we wanted. Would the planning authority consider that the neighbours' privacy would be unreasonably reduced? Would we be able to use the alley for vehicular access? Would we be able to connect to the sewer in the main road? We instructed a solicitor to produce a draft contract of sale which would be conditional on our obtaining planning permission for the house we wanted. The usual process of searches revealed some curious covenants, but no problems with obtaining water, gas and electricity on the site. Vehicular access was negotiated with the local authority and it was decided to route the drainage through the vendor's adjoining garden, past the side of his house and into the main road. Preliminary discussions with the planners confirmed that they would recommend to the planning committee that permission should be given for a single house on the site. They were familiar with the idea of self-build and timber-frame construction following their experience with the earlier Lewisham self-build projects. Meanwhile, the vendor's solicitor was being disquietingly unhelpful.

OVERLEAF *Left page, top: Through the conservatory into an established garden. Bottom: The dining end of the family room. Right page, clockwise from top left: Looking down into the family room; Playroom with contrasting ceiling height and texture; Study projects into the living-room; Diagonal bracing adds interest to the bedroom; View down from the gallery shows reused oak floor.*

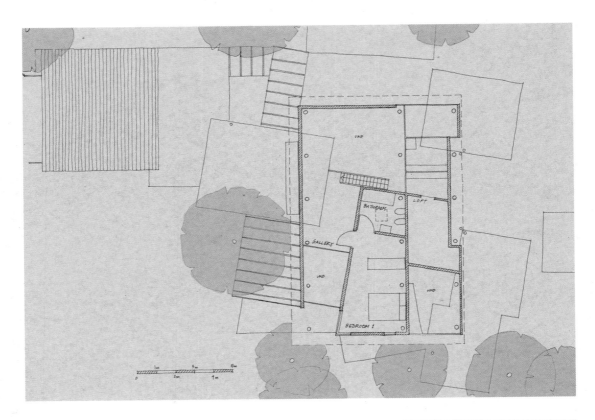

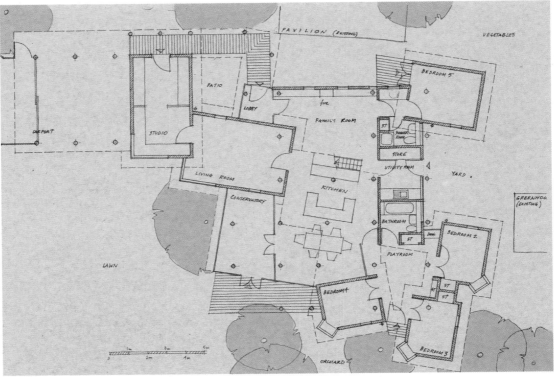

Top: First-floor plan.
Above: The plan fits around the existing trees on the site and creates a range of spaces around the house; patio, yard, lawn and orchard. The large rectangular family room is at the heart with other rooms around it at an angle.

August: the design

We had to assume that all this could be sorted out, we got on with developing a design so we could be ready to make a planning application. We were then immediately confronted with one of the biggest and most difficult questions to answer in the whole enterprise: what sort of house did we want? Anything was possible. For my part it was important that it was to be a reflection of my architectural ideas and philosophy. It would be an opportunity to extend the tradition of timber-frame self-build which has been developed by Architype, following on from the work of Walter Segal. Architype, which is a medium-sized architectural practice where I was the founding director, had been developing the Segal method to incorporate an ecological approach to building, and this would have to be a central idea. So it would be a timber-framed, low-energy house with lots of insulation, high-performance windows and low-energy lighting. It would seek to reduce the energy and resources involved in its construction by using as much locally sourced, renewable or recycled materials as possible. The choice of materials and ventilation system would seek to create a healthy internal environment as free of potentially hazardous products as possible. The use of water would be reduced, although my experience in this field was not very highly developed at that time. The position on environmental matters was a pragmatic one; the house was not going to be experimental but was going to be the best possible using tried and tested technologies with minimal cost increases. Measures had to be robust and cost-effective. Whilst we had enough money to get started, we did not have the cash to finish even a basic house of the size we wanted.

The 'patterns' developed by Christopher Alexander as a way of identifying places which have a certain, difficult-to-define quality which sets them apart from the ordinary, would be important. I was

After about a month we had a design drawn up and submitted for planning permission.

ON PREVIOUS TWO PAGES: Left page, top: Exterior. Bottom: Cladding panels are an environmental indulgence, but look great. Right page, clockwise from top left: The bathroom window looks down into the family room; Timber buildings can be built very close to trees on individual concrete pad or pile foundations; The roof is a riot of colour in spring; Family room and kitchen at the heart; The entrance lobby reduces heat loss and is full of coats and boots.

also eager to explore the forms and spaces that could be built in timber, to get away from the universality of the right angle and the constant height of rooms, to create an environment with plenty of contrasting textures and colours to engage the eye and the mind.

The planning authority would be concerned about those matters that affect the neighbours and the community as a whole: road access and parking, drainage and mains services, siting and overlooking other properties, effect on trees, appearance and materials. We needed to decide on these issues and be confident that we would not change our minds later, as we would be limited to what we could do under the planning permission that would be granted. We could leave other decisions concerning internal planning and finishes until later. It was at this stage that the major decisions that would set the way forward had to be made. The house would face out into the garden with plenty of glazing on the south; but be organized so that there was shading that would prevent overheating in summer. We would have a conservatory, positioned so that we could take advantage of the shading afforded in the summer by an existing large apple tree. The bathroom, shower room and utility room would be on the north side, with small windows. The roof would sweep down on this side of the building to reduce exposure and provide shelter from the weather. The house would be built on individual concrete pad foundations without a site slab, to minimize the damage to the existing trees, some of which were only a couple of feet from the building. (Contrary to current thinking by many Building Inspectors, structural engineers, developers, planners and others it is perfectly possible to build very close to existing trees providing that the foundations are designed to avoid problems that result from tree roots.) The house would be clad with a combination of timber, which is an economical and low-environmental-impact solution, and brightly coloured panels made from paper laminated with resin, to produce a mainte-

nance-free but relatively high-cost, high-environmental-impact board which was a financial and environmental indulgence – but it looks great on the finished building. We planned a carport for two cars, but this has never been used for its intended purpose and now houses a table tennis table.

The plan developed relatively quickly with a big family room at the heart. This contained the kitchen as well as family living space and the eating area. A children's area contained three children's bedrooms opening off a play area. A separate wing contained a living-room, and a study at the extremity shielded from noise and disruption. We would have a bedroom and bathroom upstairs, with a gallery overlooking the family room which could be used for doing things whilst still being in touch with the hurly-burly below. A guest room for visitors or live-in nanny was to be in a separate wing with a side entrance and its own shower room and kitchenette. Rona works from home, so the study was near the entrance, with a separate door so that her business visitors could come and go without having to enter the residential part of the house. The family room was to look out over the garden, with a way out onto a veranda through a conservatory. The front door was to lead into a glazed lobby off the family room to reduce heat loss and provide a transition space as you enter. The playroom had a separate garden entrance. A utility room opened off the kitchen with a door to a back yard and there was to be a large storage loft upstairs.

The house was to be sited on that quadrant of the site occupied by a now overgrown vegetable patch. The plan was devised to be built around the existing trees on the site. The form of the building centres on a rectangular roof over the family room, supported on tree-trunk columns. The single storey wings are inserted under the edge of this roof. These 'boxes' are clad where they appear inside, in the same materials as on the outside. This, together with the views out into the garden and the tree-trunk columns echoing the trees outside, is intended to reduce the distinction between the inside and the outside. The main rectangle is clad in timber boarding with contrasting panel cladding to the single-storey boxes. One of the implications of the plan as it developed is that it is not at all regular; which is good in some ways – it forms a series of defined external spaces, an entrance court, a workspace in front of the shed and a back yard for hanging out the washing – and bad in others because it increases the area of the external wall, which in turn increases heat loss, and there are many junctions which have proved to be difficult to make airtight.

The various roofs are covered in turf. A living roof has a number of benefits; in our context it softened the visual impact of the new building from the point of view of the neighbours. In a city it is particularly appropriate as the high proportion of hard roof and paved surfaces reduces the ability of plants and trees to moderate the environment, filter dust, reduce temperatures and absorb pollution. A green roof creates new habitats, which studies have shown support varied wildlife. Hard surfaces cause a high level of storm-water runoff which cannot easily be absorbed in the ground or channelled to watercourses. This leads to high infrastructure costs for storm-water sewers and problems disposing of often heavily polluted storm water. This is a particular problem in London, which has no separate storm-water system, so if there is heavy rain, the treatment works are unable to cope with the high flow, and discharge untreated sewage into the Thames. There are a number of decisions that have to be taken: is it to be a so-called intensive green roof (expensive), more like a roof garden with plenty of soil and large plants? Or is it extensive, less of a garden and more of a field or a lawn? Is it to be based on soil which is heavy especially when wet, or on one of the proprietary lightweight systems (also expensive) that are based on plants rooted in a plastic mat? Will it incorporate some means of retaining water in a mat or in a profiled underlining (which costs more and is not effective during extended dry periods)? Will it

be seeded (which can achieve a very varied selection of plants) or turfed (which will not, unless you form pockets and insert plugs – small plants – which can compete with the grass)? We opted for an extensive roof of wildflower and grass seed in soil. This would be economical and have a variety of wild flowers in spring and would dry out in summer and look brown and shaggy. The roof structure would have to carry the extra weight.

As the layout was developed it was necessary to check the practical aspects of the design. Would the structure work? What would the implications be on energy consumption? What roof pitch is possible with a grass roof? We sought advice on these aspects. This design period was one of very intensive work in the evenings.

After about a month we had a design drawn up and submitted for planning permission. The pressure to make a planning application and determine whether we would be able to proceed with acquiring the site limited second thoughts during the design stage.

November: planning permission

It was necessary for the conditional contracts for the purchase of the site to have been exchanged before the planning meeting to guard against us obtaining permission and thus making the site far more valuable without any obligation on the vendor to sell to us at the agreed price. An element of drama was brought to the proceedings because when the day of the meeting came, the vendor's solicitor had still not exchanged despite promises to do so. It ended up with me going to the meeting from work still without confirmation that the exchange had taken place. I had to phone home to check if the deed had been done whilst I was on the train. If not, I was to ask the chair of the committee to use his discretion and withdraw the item. As it was, the vendor's solicitor was contacted at home by the vendor and instructed to exchange forthwith. Our

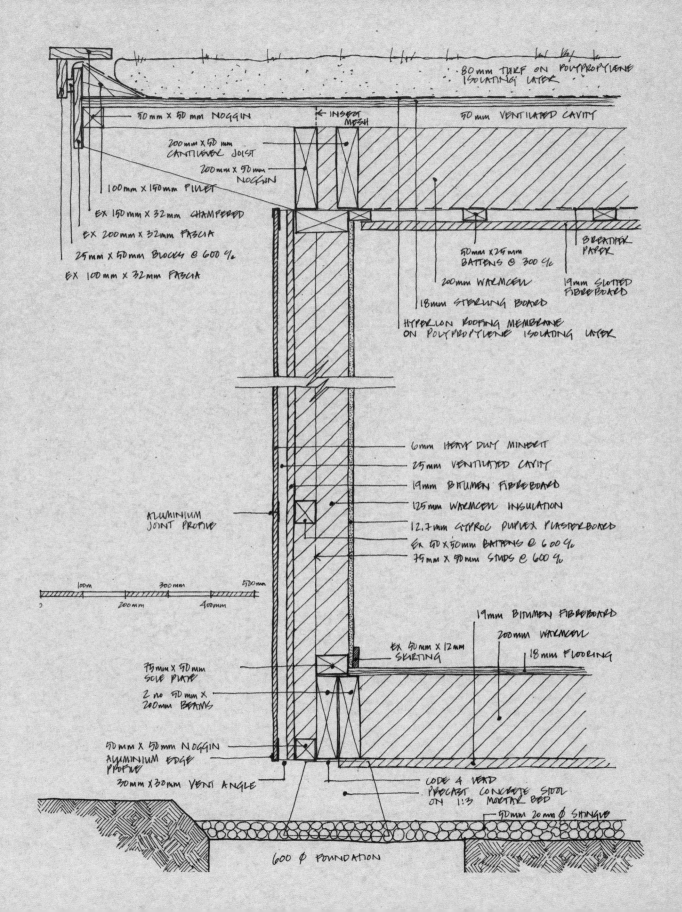

80mm TURF ON POLYPROPYLENE ISOLATING LAYER

50mm × 50mm NOGGIN

INSECT MESH

50mm VENTILATED CAVITY

200mm × 50mm CANTILEVER JOIST

200mm × 50mm NOGGIN

100mm × 150mm PLATE

EX 150mm × 32mm CHAMFERED

EX 200mm × 32mm FASCIA

25mm × 50mm BLOCKS @ 600 %

EX 100mm × 32mm FASCIA

50mm × 25mm BATTENS @ 300 %

200mm WARMCEL

18mm STERLING BOARD

HYPERION ROOFING MEMBRANE ON POLYPROPYLENE ISOLATING LAYER

BREATHER PAPER

19mm SLOTTED FIBREBOARD

6mm HEAVY DUTY MINIGRIT

25mm VENTILATED CAVITY

19mm BITUMEN FIBREBOARD

125mm WARMCEL INSULATION

12.7mm GYPROC DUPLEX PLASTERBOARD

EX 50 × 50mm BATTENS @ 600 %

75mm × 50mm STUDS @ 600 %

ALUMINIUM JOINT PROFILE

| 100mm | 300mm | 500mm |
| 200mm | 400mm | |

19mm BITUMEN FIBREBOARD

200mm WARMCEL

18mm FLOORING

EX 50mm × 12mm SKIRTING

75mm × 50mm SOLE PLATE

2 no 50mm × 200mm BEAMS

50mm × 50mm NOGGIN

ALUMINIUM EDGE PROFILE

30mm × 30mm VENT ANGLE

CODE 4 LEAD

PRECAST CONCRETE STOOL ON 1:3 MORTAR BED

50mm 20mm ∅ SHINGLE

600 ∅ FOUNDATION

solicitor was able to relay this information in time for us to proceed at the meeting. The committee granted permission to the disappointment of the objectors.

So we had the planning permission necessary for the purchase of the site to proceed.

January 1992: the model

The design was developed in the evenings over the next few weeks, and I built a card model at a scale of 1 to 20. One of the most characteristic features of the design was introduced at this stage: the curve to the main roof. It arose out of the need to reduce the height of the roof generally whilst still keeping enough headroom in the upstairs bathroom. I was very pleased to adopt this more organic approach, although later it did prove to be relatively complicated to build. In deference to the neighbours, it was also decided to retain the existing shed on the site instead of building a new one attached to the house. This was a good decision which saved time and money, was very useful during the construction and retained a link with the history of the area. The model proved very useful in visualizing the spaces in the design. The finished house is very similar to the model although a number of elements were changed in the course of construction taking advantage of one of the great benefits of self-build: the ability to change your mind or to make it up as you go along to some extent.

May: Building Regulations

The next few months saw intensive evening work developing the design, specifying materials and preparing drawings and calculations for the Building Regulations submission. I used my self-taught knowledge to calculate the structure, with the exception of the structure to the main roof. This was beyond my capabilities on two counts. Firstly,

Detailed section showing timber-frame breathing construction raised above the ground with a 'cold' roof with planting.

the methods for calculating the strength of timber poles are not covered in any of the British codes, although they are in New Zealand and Australia, where this form of construction is relatively common. This data was obtained and used for this house. Secondly, the structure of braced columns is what is termed an indeterminate structure. This means that it is not capable of analysis using straightforward calculation and has to handled by a computer, so I commissioned a professional structural engineer to carry out this part of the analysis. It consisted of around 100 pages of computer data which with my 40 pages or so of beam and column calculations formed the structural submission.

The timber used throughout is British-grown Douglas fir. This is relatively strong and durable. Locally grown timber was specified to reduce the energy embodied in the construction (imported timber requires more fuel to bring it from Canada or Scandinavia), and to encourage the use of UK-grown timber for structural purposes. This adds value to the timber and encourages replanting and proper management of British woodlands. Although timber is a renewable resource, the environmental damage caused by logging and poor forestry management and the loss of biodiversity from loss of tropical forests and mono-cultural plantation forestry are all causes for concern. One should only use timber from a sustainably managed forest, and the only way of guaranteeing that is to use certified timber.

It can be difficult to avoid uncertified timber. In particular, a great deal of plywood, especially that from Indonesia and other parts of the Far East, is made from endangered tropical forests. This ply turns up in all sorts of manufactured components such as doors and stairs. Stair strings are often made from Parana pine because it is obtainable in wide boards, but it is an endangered species. Door lippings are often tropical hardwood of uncertain provenance. It is now possible to obtain certified chipboard, ply and other building boards.

The decisions on the wall, roof and floor construction and levels and type of insulation, size and specification of windows, heating and ventilation systems were all taken at this stage.

One particular issue is how to deal with the risk of condensation. As buildings become more highly insulated and designers become aware of the need to control ventilation and air leakage to improve energy efficiency, the control of moisture in buildings becomes more critical if condensation is to be avoided. The general approach in timber construction has been to incorporate a vapour control layer, commonly polythene, on the warm side of the insulation. More recently, an alternative approach has been to develop a building envelope which 'breathes' and which is relatively porous to moisture vapour as an alternative to wrapping the building in a plastic sheet. This is going back to earlier building methods which tended to use relatively absorbent materials such as soft brick and lime mortar and lime plaster instead of heavily fired bricks set in portland cement mortar with gypsum plaster. Vapour-balanced or 'breathing' construction reduces the risk of condensation because any moisture vapour in the construction is driven towards the outside where it evaporates. A 5:1 ratio of permeability between the inside and the outside ensures that there is no risk of condensation. A hygroscopic insulation material such as lamb's wool or cellulose fibre insulation fills the voids in the construction and forms a complete insulation layer without gaps – unlike insulation batts or boards, which are difficult to install tightly between studs. Breathing insulation materials also have environmental advantages, as they are made from waste material with very low embodied energy.

The Building Regulations submission was completed by a specification which included calculations of heat losses, window areas and a condensation prediction to justify the vapour-balanced construction prepared by the company that supplies the recycled newspaper insulation which was to be used.

The amendments to the design at this stage – the roof form, the shape of the conservatory, the angle of one of the bedrooms to fit around the trees better and the retention of the existing shed – required that a revised planning application be lodged.

Whilst this information was being considered by the local authority, site clearance was completed and a start was made on the foundations with the agreement of the Building Inspector.

June: start on site

This was a hectic period, organizing tools, insurance, temporary telephone, water and electricity, a programme for the work, schedules for first deliveries of timber, materials for temporary gates and fencing, storage arrangements for materials, a bank account for payments and all the other organizational matters necessary. I negotiated with a carpenter, Steve Archbutt, one of the self-builders from the second Lewisham Self-Build project, to work on the job. I started working four days a week at Architype, and building on site for the other three days a week.

We created a new access with temporary gates and ramp formed of hardcore, and repaired the shed. This caused Steve to suffer the first and only bad industrial accident on the job, he sawed the tip of his finger off and had to be rushed to the hospital by one of the neighbours. Fortunately it healed remarkably quickly without permanent damage. Relations with some of the other neighbours got off to a bad start, with complaints about lorries coming to the site with deliveries, but others proved very helpful after the earlier objections, and one neighbour provided us with lunch every day.

July: foundations

I spent more late nights preparing a setting-out drawing for the foundations. The setting-out was

Foundation pads are excavated by a machine designed for drilling holes for fence posts.

quite complex, given the irregular shape of the building. Setting-out with pegs and string was not helped by our Alex, now a year and a half, pulling the string lines down.

I arranged for a JCB to remove the topsoil from the area to be built on, and was reminded of the destructive power of such a machine working in a restricted area up against existing trees. I organized the hire of a post-hole auger to drill out the foundation holes. This was a small hydraulic machine which can drill a two-foot diameter hole up to four feet deep which is then filled with mass concrete to form the foundation. The sixty or so foundations took about two weeks to dig using this machine, with much of the time taken removing the spoil by wheelbarrow. Some of the foundations were required to be more than two feet in diameter, and these were undercut by hand with a spade. One might consider using short-bored piles drilled using a small piling machine, as there are enough piles to spread the setting-up cost so that it becomes an economical way of doing it. One might also consider the recently developed screw-piles. These are steel screws just like a giant corkscrew which are screwed into the ground by a machine. One big advantage is that they do not produce quantities of spoil to be disposed of, and nor do they involve any concrete. The use of cement is now the single largest source of CO_2 emissions in the world.

The concrete was made on site with a mixer, and it took just over a week of hard physical graft to fill all the sixty holes. I hired a labourer friend of Steve's to work with him on the foundations on days when I was not on site. At weekends we had a great deal of help from a few friends who worked at clearing, digging and setting out. The foundations were completed by setting a paving slab on a mortar bed where there were to be timber posts. These slabs were levelled in using a dumpy level.

August: drains and services

The best environmental option for waste disposal is to connect to the main drainage system if you can (it is maintained and monitored to a high level), unless you go for a dry composting system. We decided not to. The trenches were hand dug in a period of high rainfall and the site was reduced to a sea of mud.

Plastic drainage, which is very easy to lay, was used.

Whilst PVC drainage is lightweight and easy to cut, clay drains are much better environmentally. PVC is one of the worst building materials from an environmental point of view, and we must wean ourselves off it as soon as possible. Fortunately there are perfectly good alternatives for most uses of PVC in building and it is not too difficult to build a PVC-free house, although some substitute products such as linoleum (which is made largely from chalk and linseed oil) instead of PVC floor sheeting do cost more. I have to confess that PVC drainpipes are not the only PVC used in this house, as we also used it for roofing membrane, cable insulation, rainwater downpipes and soil drainage; but more of that later.

At that time, photovoltaic panels for generating electricity from the sun were even more expensive than they are now, and a sheltered garden in inner London is never going to be a good site for a windmill. We have reduced our consumption but we rely on the grid for power (we have switched over to a green supplier).

We are supplied with mains natural gas, which is the cleanest and most convenient fossil fuel, but fifteen years later, supplies from the North Sea are running out and the worry is that we may not be able to secure an affordable and secure alternative supply. Again, we have reduced our consumption and can do so more when we fit the solar hot water panel that was planned at the outset but which has been 'this year's project' ever since.

We are connected to the water main. Our efforts to reduce our demand for water were not as far-reaching at the time as they would be now. We have spray taps and a shower which we use a lot of the time, but we did not fit low-water-use WCs, which are the most effective demand-reduction measure. One of our thirsty pans has subsequently needed replacement, which has been with a Swedish water-efficient type with a dual-flush cistern using only uses 2 or 4 litres unlike the normal 7.5 litres of a British cistern. The suppliers have recently discontinued the dual-flush cistern in Britain and replaced it with a single 4.5 litre model which uses a siphon rather than the flush valve of the continental cisterns (which although efficient at the outset will almost certainly start to leak at some point in their life – at which point they are very inefficient).

We decided not to install a water-recycling system that would collect water from the bath and shower and use it for flushing the loo, as my experience at the time on another project had been that they were not reliable or cost-effective. Similarly, our arrangements for harvesting rainwater are very basic, consisting of water butts to collect water for the garden. Our judgement at the time was that large storage tanks with pumps and sand, and ultra-violet filters for drinking-water quality, would be very expensive.

Ducts were laid for the electric main and telephone and a polyethylene water main was laid in. (This is a relatively low energy plastic with benign properties.) Ballast was laid over the whole area of the building to prevent weed growth under the house. This also provided a good base for building on.

September: timber delivered

The timber delivery was delayed because I found it difficult, working in the evenings, to produce the schedule in time. The order in the end coincided with the summer shut down at the sawmill in the Welsh borders. The timber was ordered in one lot for the whole job and was of very good quality with hardly any pieces with splits or twists that rendered it unusable. Much of the material was also in long lengths (over 7 metres) and large sections, 63 x 250mm. The timber was kiln-dried and the poles were specified to be debarked by hand with a maximum 10% taper in the length.

Douglas fir is a moderately durable species of timber, and so no timber treatment has been used on the job. Such treatments are generally toxic, and should be avoided if possible. It has been suggested

Above: UK-grown Douglas fir timber in a rack.
Below: Friends rally round on Saturday morning to raise the frames.

that the prevalence of timber treatments in Britain is the result of very successful marketing by the chemical industry supported by the insurance companies, building societies and poorly trained surveyors. There is nothing like the same reliance on timber treatments in North America. Timber is generally not at risk if it does not stay permanently damp. It can get wet so long as it can dry out again afterwards. The end grain is vulnerable, and it is always better to leave a 3mm space around the end of timber rather than butting it up tight in the hope of keeping water out. Use large roof overhangs, don't build timbers into damp walls, set windows and other timber details back from the face of the building, and avoid sapwood.

The timber comprised two large articulated lorry-loads. They looked huge from the ground – the first and largest load took six men four hours of very hard work to unload and get into the specially erected scaffolding rack. This stored the timber under cover until the main roof was on in December when the rack was dismantled and the scaffolding returned off hire. It enabled the different types of timber for columns, beams, joists and cladding to be racked separately for easy access. I sorted the timber and found, inevitably, that some pieces had been wrongly delivered. I arranged for them to be changed. A considerable amount of time involved in building seems to be spent on checking and reordering materials.

October: framing

Another stint of working late in the evenings was needed to produce drawings showing the key dimensions for making the frames. I also produced detailed drawings for the fabrication of the galvanized steel plates that were necessary to bolt the beams to the posts. The six main frames were made flat on the ground and lifted up into position. Each frame consisted of three tree-trunk columns with double beams at the roof and floor levels, which were bolted together with braces at roof and

Above: A beam is offered up to be marked for cutting at an angle and drilling in the correct position for bolts. Below: The roof membrane is sealed by melting with a hot-air gun, and is glued down to the discs in the foreground to prevent the wind lifting the membrane, whilst still allowing it to move horizontally.

floor levels. The tree-trunk columns had to be notched for the roof and floor beams. Because the poles were irregular we stuck stringlines for a reference point on each log. We devised a gauge to ensure that the seatings cut into the logs for the beams were parallel and the correct distance apart. We made one or two mistakes, drilling the wrong size hole for bolts for instance, and they took time to overcome. We suffered our first burglary; the shed was broken into at night and all the power tools were taken. I replaced the tools and bought a second-hand tool vault. The people at the tool shop said that they had workmen in every day who had lost the lot. The insurance covered the loss.

The frames were raised into position with the help of friends on three Saturday mornings. This required much phoning around to assemble the required number of people at the same time. This was a moment of great uncertainty: would they be too heavy, and would there be a danger if they fell? It took twelve people to lift each frame into position using pushers bolted to the roof beam fixing plates at the top. We had a heavy rope belayed around a tree in case of emergency, but as it was they went up in time-honoured fashion without any difficulty. The last two frames had to be carried in vertically in two parts because there was not enough room to raise them whole.

Once the frames were up, we adjusted them for position and verticality, and positioned the DPC at the base. One refinement over Walter Segal's original detail was to use bitumen-coated waterproofing membrane cut to the shape of each post in addition to the normal lead detail. The bitumen is squeezed into the end grain of the timber and gives a better seal at this vulnerable position. Also, where the posts were external to the completed building and therefore subject to regular wetting, I inserted a 50mm upstand between the post and the foundation so that the post was not actually standing in the wet. This upstand was made out of a 50mm slice of large diameter plastic drainpipe filled with a strong mortar.

November: beams and joists

I spent more late evenings doing a dimensioned roof layout showing the position of the rooflight and noggings which would support the top of partitions later. Double beams were fixed between frames at roof and floor levels – a heavy and difficult job when at roof level. They had to be lifted 6m by one person at each end on a ladder, offered into position, cramped and marked for length with the correct angle for the cuts. They were lifted back down for cutting, then back up, cramped into position, drilled for bolts using metal plates as a template and bolted in position. A lot of time was spent manoeuvring ladders into a position for drilling that was comfortable. We did not use any scaffolding during the job but constructed working platforms when necessary.

This was the time of my nastiest moment on the job: the drill jammed in the 'on' position, and rather than be thrown from the ladder I let go and it wrapped the cable around my arm. I managed to turn it off, but it was a frightening and painful moment – twenty feet in the air, and not helped by it being very cold and rainy at the time. The roof joists were heavy to get into position, being 50 x 250mm in section and 6m long. The ends of the joists to the rear slope of the roof needed a very deep notch to sit in the joist hangers, and I had to check the shear load. I employed Steve's friend to help him when I was not on site during the week for this part of the work, and I got help from friends at the weekend. The beams and joists at floor level were easy after this. I had my first day off, as I had flu.

December: main roof

The grass roofs require a membrane roof, rather than a pitched roof with tiles or slates.

These roofs are all designed as flat or sloping, and are so-called 'cold roofs'; that is, the deck supporting the roof membrane is on the cold side of the

insulation. In cold roof construction it is good practice to fit a vapour control layer on the inside of the insulation. However, this merely reduces the amount of vapour that migrates outwards through the construction towards the underside of the roofing membrane and deck, where there is a ventilation space open to the air at the edges of the roof to allow cross-ventilation, which carries the moisture vapour away. The vapour control layer does not have to be perfectly sealed, because there is ventilation, and so the construction is relatively risk-free – unlike a warm roof construction, where the vapour control layer does have to be well sealed as otherwise the risk of condensation is high. Nevertheless, in a cold roof it may be that the design relies on the vapour control layer to double up as an air retarding layer to achieve airtight construction, in which case it will have to be well sealed.

The roof construction of this house is part of the 'breathing' or vapour-balanced envelope discussed above. The construction consists from the inside of a ceiling of 100x25mm square-edge Douglas fir boards fixed to 50x25mm battens fixed to the 50x250mm joists on top of a building paper which incorporates a microscopic polyethylene film to resist the passage of moisture vapour. This is stapled below the joists, and the space between the joists is fully filled with cellulose fibre insulation retained at the top by netting staples over the joists. 50x50mm battens fixed on top of the joists form the ventilation slot, which is protected from insects and birds

by mesh at the edge of the roof. 100x25mm square-edge softwood boards cut to fit the double curvature of the roof form the deck with the membrane and ballast on top.

We fixed fascias using a string line to get a straight line on a structure which is all at angles. I had to swap material around to get enough depth to the fascia at the front. The Warmcell recycled newspaper insulation was delivered in compressed bales, with a blowing machine on free loan. I carried out a trial section in the main roof to check the viability of blowing in insulation from below once the roof deck was fixed. In this way the insulation would not get wet. It appeared to be possible, so I proceeded to fix the roof.

We then lost our first and only day due to weather. We stapled PVC ventilation angles to the edges of the roof to keep insects out of the ventilation voids, but subsequently discovered wasps passing through these vent angles and so abandoned this product for the ventilation voids in the external walls in favour of much cheaper aluminium woven insect mesh. We fixed the roof boarding. Fitting the boards to the curved roof felt more like boat-building than carpentry.

I took time off work but had to go in a couple of times to deal with crises – business was not good, so we had to make some redundancies. We fixed the single-ply PVC roofing. I used a PVC membrane as it was from the only supplier that I was aware of that would supply the material and let you fix it yourself. Fixing it myself meant that I could do the various roofs at a time that suited me, and of course it saved a good deal of money. I have learned recently of a Thermoplastic Polyolefin (TPO) membrane that has a better environmental performance than PVC, and the manufacturer will supply it for you to fix yourself providing you go on a training course.

The single-ply membrane roof retains the original Segal idea of a loose-lay membrane which is not fixed to the roof but which is draped over it in the manner of a tablecloth and prevented from being sucked off by the wind by weighting it down. This can be with a pool of water, as in his first temporary house at the bottom of his garden in Highgate; or with gravel, as in the later Lewisham self-build houses; or with soil and planting, as we did. The edge of the roof is protected by, but not fixed to, a capping. This arrangement avoids the problems associated with flat-roof construction in Britain, where flat roofs have not been specified properly and have acquired a very bad reputation in spite of the fact that many large modern buildings – warehouses, factories and supermarkets – have flat roofs. The membrane can move with changes in temperature, and the timber deck can move with changes in moisture content from one season to another without stresses occurring where they are bonded together. The ballast on the roof protects the membrane from ultra-violet radiation, which tends to make it brittle and prone to cracking. It also keeps the membrane at an even temperature and prevents puddles forming. This is important because a puddle keeps the roof membrane cool, whereas the dry area around it can become very hot. This causes differential expansion at the edge of the puddle which can lead to weakness in the membrane, a crack and eventually a roof leak.

The PVC system has the added refinement of a washer detail that retains the ability of the membrane to move horizontally whilst keeping it anchored to the roof against uplift from the wind.

This is achieved by screwing PVC discs to the roof at 600mm centres with metal washers. The roof membrane is then glued to these discs. We used these discs on the main roof as we estimated that it would be some time before the roof was ballasted with soil during which time the roof would be vulnerable. The PVC material comes in a roll which is seamed together with a hot-air gun. The membrane is laid over an underlay and welded to a metal angle at the edges. There are a number of advantages over Segal's earlier felt roofs; the process is self-buildable and the membrane is very flexible, which is a particular advantage in this case where the membrane has to accommodate the curved shape of the roof.

The rooflight is a triple-glazed dome with insulated upstand to which the roof membrane is bonded. This is a very simple and effective detail. Roofing started during a very wet spell. The hot-air welding technique worked well in the damp, but gluing the membrane to the discs did not. I needed to dry everything out as far as possible with the hot-air gun. Roofing ended during a very cold spell. Working on the roof was tricky when it was icy as the PVC was incredibly slippery. We lost control of a roll of roofing which went over the edge and took an hour to sort out. I had my other nasty moment when I was working alone and slid down the roof out of control. Fortunately I was able to stop against the upstand at the edge.

There was no work on Christmas Day. We had a topping out ceremony on the 28th December which was attended by 65 friends who came back to our old house for drinks. The house was not big enough. It felt like a big moment and the end to a good year's work.

January 1993: floors to wings

The floor structure of single-storey wings was set out on blocks and fixed together. Setting out the angled shape proved more straightforward than anticipated. Our new baby was born in the kitchen

at Segal Close. The floor structures were supported on concrete stools cast in plastic buckets obtained from the flower stall in Catford. The procedure was to level the structure on temporary blocks, jack it up and slip the concrete stool under on a thick mortar bed. The jack was then let down so that the floor structure sank down onto the blocks, squeezing mortar out until the concrete stool was set at the right level. The blocks were removed when the mortar had gone off. This procedure worked very well. Rona and the new baby were admitted to hospital for ten days when she was three days old. I worked on site with our two-year-old, Alex, in his pushchair. I could work in this way for an hour or two before Alex got bored. I was able to work out how to proceed sufficiently to allow Steve, our carpenter, to get on with the job.

February: walls and roofs to wings

The stud wall panels were nailed together using the floor structure as a base, and erected in turn. Work proceeded quickly. Problems occurred as usual as soon as a more complicated part of the construction was encountered, in this case the framing for the bay windows. It proved very difficult to set out the structure in mid-air. This was to cause some problems later, as we had to adjust subsequent work to take account of the discrepancies in the setting out and level of the framing. Steve phoned to say that there was a problem so I looked in on the site on my way to work for the first time. It was a mistake in cutting the taper to the overhang to the living room roof – but nobody will ever notice.

The sequence for the smaller flat roofs on the wings was to put in the insulation from above, rather than below as we had done on the main roof. The procedure was as follows: we stapled vapour control paper to the underside of the roof joists; 50x25mm battens which will eventually support ceiling boards were nailed to the underside of the joists; we blew in insulation from above to fully fill the depth of the joists, which was 200mm; we fixed 50x50mm battens across the joists to form the ventilation gap, and then fixed the OSB (Oriented Strand Board) deck. OSB is similar to chipboard, but made from larger flakes of wood rather than sawdust, and used for structural purposes in place of plywood which is much more expensive. It does contain a moderate proportion of fossil-fuel-based resin to bind it together, but it is made from very low-grade timber and timber waste and it is so much quicker, cheaper and better than tongued and grooved boarding – another example of the compromises that are necessary. Finally, we fixed the edge fillets and the metal angle, and loose-laid the roof covering without discs to hold it down. We had hoped to be able to do all that for a small roof in a day, but this proved impossible. We were able to get the deck fixed in a day, at which point temporary plastic sheeting protected the insulation until the next day.

I misjudged the fine line between wasting money by ordering too much and wasting time by having to order more later. I had to reorder the vent angle, but was able to obtain some surplus metal angle from the nearby self-build job under construction. From now on we were working under cover.

March: external walls

This was a period of intensive decision-making on site. Should we have a full height window slot at both ends of the west wall? How was the conservatory roof going to be supported on the front wall to the kitchen? Would 16mm diameter steel studding be strong enough to support the front wall to the kitchen? There were a myriad other questions.

We fixed 50x50mm battens to the external walls. These formed the outer layer of a double layer wall construction which gave a cavity of 125mm to be filled with insulation, avoiding cold bridging at the studs. It also meant that the wall was constructed of smaller sections of timber which were cheaper and easier to handle. Work proceeded rapidly but, as usual, time was needed to puzzle out what to do at awkward junctions.

We nailed bitumen-impregnated softboard to the outside of the external walls. The permeable softboard formed the outside layer of the breathing wall construction. The boards are tongued and grooved on all four edges, which meant that they could be fixed without having noggings under all the edges. This saved a lot of time. The inner vapour-check of the wall construction would be incorporated as a very thin plastic film applied to the back of the so-called Duplex plasterboard.

We carried soil up onto the roof. I had sought the advice of an ecological landscape designer, who advised that the topsoil that I had set aside from the foundation excavations was far too rich to support vegetation on a roof. The drought-resistant species that can survive through a long hot summer need a very impoverished soil: too rich a soil encourages the growth of lush green species that die as soon as it gets dry. We ordered chalk and ballast, which were mixed with the sieved topsoil in varying proportions to give different soil types on the different roofs of the house. The back slope of the main roof is relatively steep (about 30⁰) and this has a framework of battens laid on the roof which supports turfs from the garden laid upside down. The root mat in the turf prevents the soil being washed away, and the battens stop the turfs sliding down the roof. The soil mixture was carried up ladders onto the roof in buckets and spread to a thickness of 80mm on top of an underlay to protect the roof membrane. There are 50 tons of soil on the roof! We had help from friends at the weekend and I hired Steve's friend during the week to help with the labouring again. The roofs were seeded with grass-seed mixtures with some wild flower seeds mixed in with them, the species being selected to suit the particular soil type on that roof. There is no moisture-retaining layer or drainage layer in the build-up of

FACING PAGE Top: Timber stud wall panels are simply nailed together and fixed in position. The diagonals resist wind loads. Middle: Recycled newspaper insulation is blown into the roof on top of blue vapour-control paper. 50x50mm battens form a ventilation gap below the OSB roof deck. Bottom: Horizontal battens nailed to the outside of the studs eliminate cold bridging. Battens support blue vapour-control paper on the ceiling. Fibreboard with T&G 4 edges fixed to outside of battens with no need for noggings to support edges.

the roof, so the vegetation experiences extreme conditions – from waterlogged to completely desiccated. It is anticipated that the vegetation will take three seasons to become established. The effect is very colourful in the spring with the wild flowers, but the roofs dry out completely in summer. The growth has been insufficient in some places to stabilize the soil on steep slopes, and erosion prevention measures have had to be implemented. The soil that had washed away has been replaced, re-seeded and covered with a biodegradable jute fabric through which the plants can grow. This has been successful in preventing further trouble, and the jute has rotted away now that the vegetation is established.

At this point Steve decided that he was going to find his fortune working in Germany. It was a sad loss as we had worked well together. Steve was tremendously energetic, with a great drive to get things done. He recommended a carpenter friend, Noel Gaskell, who lives locally. Noel visited the site and it was agreed that he would take the job.

April: roof fascias

We took Easter week off for a family holiday in Wales. On our return we screwed roof fascias and cappings in place. It took a surprising amount of time: sanding, painting, drying, measuring, marking, cutting, drilling, notching and fixing.

May: insulate main roof

The Warmcell insulation to the main roof was blown in from below. First vapour control paper was stapled to the underside of the joists, then the insulation was blown in and battens were nailed to the underside of the joists ready to take the ceiling boards. Temporary staging was built over the large voids to gain access to the ceiling.

June: first fix plumbing

I spent more late nights working out the plumbing. The headroom in the loft was very tight for a gravity-fed hot water system. However, it looked as if it was possible, and so this was the basis for the design of the pipework. I sought advice from an energy consultant friend, John Willoughby, and the heating system was designed with two zones, one for the living spaces and one for the bedrooms, and two heating circuits were installed accordingly. The study has its own gas convector independent of the rest of the house; it has proved useful to have a very responsive heat source in the office when the rest of the house is not in use, although I do not think the gas convector is very efficient. Hot water was to be stored, rather than heated by a combination boiler which heats hot water as and when you need it, and a solar hot water panel was to be fitted.

I had to fix the size of all the windows so that the heat losses could be calculated. Then the radiators could be sized so that the pipework could be installed in the right place. This meant having a complicated discussion with Rona about ventilation. She wanted a small opening window in each room for ventilation at night in addition to the main opening window. The house was designed with a minimum of opening windows, and relied on a controlled ventilation system. Because of the layout of the house, this could not be a passive system but had to rely on a fan drawing air out of rooms that are a source of moisture, such as kitchen and bathrooms. The rate of extraction is controlled by a humidity-sensitive extract register in the ceiling. Fresh air is drawn from outside the building through humidity-controlled inlets in each of the habitable rooms in the house. The inlets admit more air if the air in the room is moist, if the room is occupied for instance. A compromise was reached whereby small openable vents would be provided in each of the ground floor bedrooms in addition to the mechanical extract system. This

Above: The grid of battens prevents soil sliding down pitched roof.
Below: The green roof makes the house fit into its surroundings and minimizes its impact on the neighbours.

enabled work to proceed with the plumbing. The wall construction was modified to accept the additional vents, and they were manufactured along with the rest of the windows. They did look rather unnecessary when they actually arrived on site, and so the wall construction was modified back again. This was fortunately one of the very few bits of work that had to be done again throughout the job. In practice the mechanical system has proved very good. It provides fresh air continuously day and night and so you do not get the sensation of insufficient fresh air. The system installed does not have heat recovery (which transfers heat from outgoing stale air to incoming fresh air by way of a heat exchanger, thus reducing the heat lost in ventilation). The design of heat exchangers now means that they are relatively efficient and cheap to produce. It is important, however, to ensure that the power rating of the fan is as low as possible otherwise you can lose what you gain in reclaimed heat to using additional electrical energy, which leads to greater emissions. An efficient system would run on a fan rated at 6W – unlike the fan in the system that I fitted, which is rated at 40W. A heat-recovery system requires ducts to supply the pre-heated fresh air as well as ducts to extract the stale warm air, and these two duct systems can be difficult to accommodate in the building design. A supply system does allow you to filter the incoming air, which will improve air quality in the house, but you will need to clean or replace the filters every year.

The pipework in the house is a plastic push-fit system which is absolutely marvellous: it is quick to make the joints, many fewer fittings are necessary because the pipe is flexible, and the system is foolproof (of all the hundreds of joints in the building, only two dribbled slightly when the system was tested and they were both on conventional plumbing fittings. Not one of the plastic joints leaked a drop, which is very impressive.)

In this house all the pipes and wires are buried in the wall, roof and floor construction. This can lead

Above: Humidity-controlled fresh air inlet above window.
Below: Reused oak flooring.

to problems maintaining or adapting systems. More recently I have generally designed walls with a service void formed between the plasterboard lining and a separate paper vapour-check membrane. This service zone can be formed in various ways, but a simple solution is to staple the vapour-check membrane to the inside of the studs, then fix horizontal battens to which the plasterboard is fixed. This enables all the pipes and wires to be routed around the building without penetrating the vapour-check membrane which is also providing airtightness. It also permits relatively easy access to the services for maintenance and alteration or upgrading. The service zone can be insulated to improve the performance of the wall.

I had to have week off with a virus infection, and felt exhausted afterwards. I never used to be sick like this! I also had to spend a weekend at the office to meet a deadline on a project. Maybe I was just trying to do too much. I needed to get the window details sorted out so that I could order the windows which were now overdue – more late nights. I constructed the bay window structure, and did other preparatory work for the windows so that I could measure the openings and order the windows. Other issues that had to be sorted out for the window quotation included the locking arrangement, whether laminated glass was required for security and whether any obscured glazing was needed. We had now been a year on site. Progress had been good generally, but we had never recouped the time it took to sort out the foundations at the beginning and there had been cumulative delays since. These totalled around two months at this stage.

July: first fix wiring

I spent yet more late nights working out the electrics. The power supply to the bedroom areas is arranged as spur circuits. This costs a little more than the usual ring circuits, but has the advantage that it avoids the problem of very high electromagnetic fields (EMFs) being generated if there is a bad connection in a ring circuit. It is unclear what effect EMFs have on the human body, but it has been suggested that they are harmful if you experience long exposure such as during sleep. To fully guard against this potential hazard you should use (expensive) screened cable. A demand switch can be fitted that switches the power supply off automatically when there is no demand at night (but make sure the freezer has a separate continuous supply).

I had a terrible time trying to decide on light fittings: good fittings are almost inevitably imported and far too expensive. It is important to fit dedicated low-energy fittings incorporating the control gear for a low-energy lamp – do not fit conventional light fittings and use low-energy lamps incorporating their own control gear because it is much more expensive – you throw away the control gear every time, and there is always the temptation to use an ordinary tungsten bulb. In the end we fitted a good many low voltage spotlights to provide visual interest and contrast – although more efficient than conventional tungsten light bulbs, they cannot be considered low-energy lighting, for which you need Compact Fluorescent Lamps (CFLs). Fortunately there is now a full range of fittings available which take CFLs.

I checked to see whether it would be necessary to increase the size of the cables because they are running in insulation, which might cause them to heat up. The advice was that it would not be a problem. I would now specify cable with PVC-free insulation although we used standard cable in this house. The PVC-free cable is specified on large public buildings – it is much safer in fire as it does not give off large quantities of toxic smoke. It is now no longer possible to carry out your own electrical installation as it must be done by a fully qualified electrician.

August: flooring

We nailed 22mm-thick bitumen-impregnated fibreboard up under the ground-floor joists to support the insulation which fully fills the 200mm depth of the joists. This was one of the worst jobs so far, lying on our backs in the restricted space under the building, although the T&G boards did mean that you could support one board from the edge of the last one fixed which helped. Also at this time the drainage connections were brought into the building.

We started flooring out of sequence as the intention was to have had the windows fixed and the building weathertight. I decided to go ahead without windows and put up plastic sheeting for protection. I had purchased a quantity of second-hand oak from vinegar vats from a factory that had closed down in London Bridge. The vats were kept full of water when the factory closed to stop them from collapsing so the material had a very high moisture content. I decided to go ahead anyway. The oak boards are slowly giving up their excess moisture and are shrinking as a consequence, which means that the tongued and grooved boards no longer fit tightly together, but the oak looks absolutely wonderful when polished and sealed. We had a working party with friends which was a good day and good work was done. We then had a week off in Scotland.

September: plasterboard walls

I decided which make of radiators to use, so that I could incorporate fixing noggings in the wall construction. I chose radiators with inset headers so that the pipes come into radiators unseen from behind. They look good in the finished job.

Plasterboard was fixed with screws using a battery-powered screwdriver. Tapered-edge boards were used which will have flush taped joints with filler. This technique took a while to perfect (in fact I never really perfected it) but it gave a flush finish to the walls and was probably quicker and certainly cheaper than the classic Segal-method battened wall.

I carried out a trial section of wall insulation. This was injected into the cavity in the wall from the outside through holes drilled in the fibreboard. I think on reflection that this was not the best method. Much better would have been either to staple scrim (lightweight fabric) over the inside of the studwork and inject insulation through holes in this, so that you can see exactly what is going on before fixing the plasterboard, or to use the wet spray method where the insulation is mixed with a little water and is sprayed into position rather than blown.

I had the erroneous idea that the cellulose fibre insulation would provide an airtight building. It does not. I had the building air-tested some years later and found to my horror that at 13m³/m²/hr (that is cubic metres of air passing through one square metre of exposed envelope of the building per hour at a pressure difference of 50 Pascals between the inside and the outside), it would not even achieve the current standard under the Building Regulations of a maximum of 10m³/m²/hr – which is not a very high standard. The average for new houses is estimated to be a value of around 8, a low-energy house should be 3 – and less if you fit a Mechanical Ventilation with Heat Recovery (MVHR) system. Positive measures need to be carried out to limit air infiltration. This can be done either by bedding internal floor panels and ceiling and wall plasterboards to the timber frame in mastic, or by providing a membrane on the inside which can also act as a vapour-control membrane, and/or providing a 'housewrap' membrane on the outside. Sealing around intermediate floor junctions can be difficult if it is on the inside, and connecting a membrane in the ceiling to one outside the wall can be difficult if you have a cold loft in the roof combined with an outside air-retardant membrane. One area that is critical, however, if you decide to tackle airtightness, is around service penetrations.

Four strong people lift a large double-glazed window to a high level up a purpose-made 'ladder'.

Meanwhile, I was working on a chapter in a book on Housing and the Environment in my spare time!

October: windows

The high-performance timber windows are double-glazed with low E (for emissivity) coating on one face of the cavity. The coating reflects infra-red and reduces the rate of heat loss through the glass. Now you can get what are described as 'soft' coatings which are more efficient. We chose not to have the cavity filled with argon gas as a cost-saving. Argon is more viscous and syrupy than air, and slows down the rate of transfer of heat across the cavity. Another

refinement to the glazing units is to specify 'thermally broken' spacers. Normally the spacers around the edge of a double-glazed unit are sections of aluminium. This transmits heat very effectively at the edge of the unit. You can see the effect of this in our house, where on a cold winter's morning there will be condensation formed around the perimeter of the windows where the glass is cold at the edge. Thermally broken spacers are made of two aluminium sections held together with resin, which is a relatively good insulant. Nowadays I would specify triple-glazed windows. These are either a sealed triple-glazed unit like a double-glazed unit but with an extra set of spacers and another layer of glass, or in the form of a coupled sash – a double-glazed window with an extra separate window next to it with a single sheet of glass. This type of triple-glazed window provides good sound insulation as well as thermal insulation.

We decided to buy windows made to order in this country which are a copy of a typical Scandinavian high-performance timber window and of similar cost. Unlike typical British windows they have relatively heavy sections so that they are rigid and no projecting timber sill which is liable to rot. They are fitted with an aluminium sill and aluminium bottom beads to hold the glass (which are otherwise small timber sections in a vulnerable position, and similarly liable to rot). The Scandinavian windows have some advantages over the British windows: they come with a high quality factory finish which is sprayed on before glazing and so goes into all the rebates; the doors are ready-hung in the frames; and both doors and windows are fitted with triple-locking systems which are very secure. They are made in fully automated factories which export all over Europe which enables them to be competitively priced. The British window is not such a good specification, being produced in a workshop which has had much less investment in equipment and so the labour cost element is much higher. The advantage of the British firm was that they were much better able to accept revisions to the order; with an automated factory, once you give the order to go you are locked into a production schedule with no room for changes.

There were over 50 windows, some of which were sloping shapes and some of which were large and high up. The largest and highest was 6 feet wide, 7 feet high, double-glazed, and weighed around 500 pounds. Its final position was to be with the top of the window 20 feet above the floor. Four of us could just about stagger around with it. The question was how to get it up into position. The method adopted was to construct a kind of giant's ladder below where the window was to be fixed. This ladder could accommodate five people who lifted the window one rung at a time. The top of the window slid against battens fixed either side and a second set of battens prevented the window from toppling backwards. The window was in place within minutes.

The tolerance gap between the window frame and the structure was stuffed with mineral wool insulation from the outside and sealed with mastic from the inside, unlike the normal practice in Britain of sealing around windows on the outside which has the danger of trapping condensation inside the joint.

November: plasterboard ceilings

I completed various plumbing and wiring bits and pieces in preparation for fixing ceilings. I filled the first floor under the bedroom with Warmcell, which also acts as good sound insulation. I fixed ventilation ducts to bathrooms and kitchen.

December: fibreboard ceilings

Slotted fibreboard panels were glued to battens in the single-storey wings. The idea was that the slots would conceal the position of joints, particularly where the panels have to be cut and fitted around the tree-trunk columns, but in fact it was very difficult to get the panel joints neat. These fibreboard panels were one of my least good ideas;

the panels are very easily damaged and do not look good with the poor joints. I decided not to fix these panels in the kitchen and wasted a couple of hundred pounds worth of material.

I also worked at painting windows, filling window fixings with timber pellets and hanging external doors and fitting ironmongery so that the building was secure for the first time.

We had been on site 18 months now. The original programme showed us moving in at this time. It felt as if progress had slowed down and a lot of time seemed to be going in fiddling around with many things at once without ever being able to finish one complete stage in the job. I felt very weary now but we had to keep plugging away as there was still a lot to do. I prepared a revised programme for completion which showed another five months work. In fact it was to be another eight months before we moved in. One of the implications of the extended construction period was that the original estimate for paying for labour had more than doubled. We decided to raise a second mortgage on our existing self-built house.

January 1994: timber ceilings

The ceiling to the main roof is square-edge Douglas fir boards nailed to battens with a gap between. This allows the boards to follow the shape of the curved roof without cutting. There was a great deal of planning to do, and Rona and I spent a day visiting kitchen, bathroom and lighting showrooms looking at fittings.

February: stairs

It became clear that the original design for the staircase would obstruct the space and view in the big kitchen. A rethink was initiated whilst standing in the space to see the effect of different arrangements. This ability to change your mind as you go along is one of the important advantages of building your own house. We decide to make the stair treads of Douglas fir plywood, which has a lovely colour and bold grain pattern. It is manufactured in Canada using glue which does not suffer from the undesirable property of giving off toxic gas into the interior of a building, unlike the more common urea formaldehyde glues.

We cleared out junk from Segal Close prior to putting the house on the market. We were somewhat apprehensive about how difficult it was going to be to sell what is a very comfortable house in a good position but nevertheless unusual in design at a time of recession in the housing market. As it was, the first estate agent we placed it with produced nothing. Our first free advertisement in the architectural press produced a number of architects who were only interested to see inside a Segal house. We were beginning to get anxious after about three months and were about to place the house with another local agent but decided to take out another free ad first.

This produced exactly what we had hoped for: someone (an architect) who walked in through the door and said, "I have been looking for an interesting house to buy for four years, and this is it." He offered the asking price, the building society surveyor valued the house at the same figure and the sale went through in about two months. This gave us a date to work towards.

March: plasterboard joints

We found it difficult to get a very good finish at first, but soon improved with practice. The random orbit sander was wonderful for sanding down the joints, but it makes an astonishing quantity of really fine dust.

April: second fix plumbing

We fixed the bathroom, the boiler (which is a high-efficiency gas condensing model) and the hot water cylinder. This has two primary coils – one from the boiler and the other from the projected solar panels. After seeking advice I decided to install a mains-pressure hot water system, although this meant adapting some of the pipework that had been installed already. This part of the work found me back at the plumber's merchants practically daily, sometimes more than once in a day, for bits.

May: second fix electrics

We ran a heavy-armoured cable under the building from the meter in the shed to the intake position in the utility room. I hired an electrician to terminate this cable at each end, which was beyond me. While he was at it he installed the consumer unit. He made the whole operation look very easy. We switched on and . . . there was no bang – it worked! A small milestone: power in the house.

The heating system is controlled by an electronic gadget which learns the thermal behaviour of the building over the period of about a week. The controller then optimizes the start-up time of the boiler to achieve the set temperature at the time required, thus obtaining maximum efficiency from the system. This takes into account the effect of the outside temperature. The difficulty is that I, like many people, find it difficult to deal with digital controls, so the system is often not running at its most efficient because I have not set the controls to match the demand – this is not entirely straightforward.

We brought the water pipe into the building. Another small milestone – water in the house – and not too many leaks! We were due to be moving in now according to the revised programme, but it was to be another three months before we were able to do so.

June: wall insulation

I injected Warmcell insulation into the wall cavity from the outside through holes in the fibreboard. I took some boards down to check that the cavity was fully filled. There were some places where voids remained, particularly at the corners of the building so I drilled more holes and topped up the insulation. I injected Warmcell into the internal walls as well to provide sound insulation.

We brought the gas into the building from the meter in the shed. My friend Terry, one of the original Lewisham Self-Builders, popped by one Saturday morning and tested the gas installation for leaks. There were none, again! I was getting quite used to things working first time by now. This was the moment to fire up the heating system – again, it worked first time!

Another of the Lewisham Self-Builders came in for a day to plaster the kitchen ceiling. The process of plastering made a great deal of mess and confirmed my prejudice against wet construction.

July: floor finish

The second-hand oak flooring was very uneven in thickness, so we used a planer to achieve a relatively even surface for the floor sander. Even so, the powerful, hired-in belt sanding machine found the oak heavy going because it was so hard. The first time we used the machine it streaked across the room out of control and crashed through the plasterboard because the fixing for the handle was not secured properly. It was a time-consuming operation sanding with successively finer grades of abrasive and applying two coats of sealer. The result however, is wonderful, with a deep, rich colour and interesting grain pattern.

The floors in the bathrooms were finished with lino and in the bedrooms with cork tiles – but cork tiles without the usual PVC facing. We were able to obtain these tiles, which are imported from Portugal, before the PVC was applied in this country. The cork was finished with two coats of organic sealer and has performed very well for the last 11 years. The floor feels very comfortable to the touch.

The kitchen units had been made by a friend who is a local cabinet maker, using the same

Colour and textures: reused oak and natural cork flooring, natural emulsion paint on the walls.

Douglas fir plywood that we used for the stair treads and a formaldehyde-free grade of MDF (Medium Density Fibreboard). They were beautifully made, and look very good, with a contrasting colour and grain pattern to the oak flooring. We fitted granite worktops, which were a bit of an indulgence, but so much better from a practical and environmental point of view than laminate-faced chipboard. The kitchen was equipped with a multi-compartment waste bin to take compost and other fractions of waste for recycling. The internal decorations were underway using organic paints and sealer, and the house was full of the delicious scent of citrus.

August: moving in

The pace became pretty hectic at this time, and the site diary has no entries for the last three weeks or so because of the lack of time. I ended up working one or two late nights on site till midnight. The cooker was fitted, and the skirtings and architraves were sanded, sealed and fitted. Rona visited the site the day before we were due to move and had serious doubts about whether it would be possible, given the state of the place. Help from a friend equipped with a broom, hoover and damp cloth turned the place from a building site to a new home. The move itself passed off remarkably smoothly, and we were installed amidst a jumble of belongings by lunchtime. A good many bottles of champagne were consumed that evening.

There were one or two drawbacks to living in an unfinished house; there were no doors, which visitors found a bit disconcerting in the toilet. There was still a great deal of work to do before the house could be thought of as finished. We decided to keep Noel on, and he worked till December fixing cladding and weatherboarding to the outside and hanging doors. It was a balancing act how long we could wait to reclaim the VAT on the job – which amounted to a few thousand pounds – as we would not be able to reclaim VAT on any material bought after that time.

I continued with the walkway to the front door, the entrance lobby, conservatory, verandas, upstairs bathroom, shower room, kitchenette, shower room, new front fence and carport. I still have the solar panel and water butts to fix eleven years later. And then there is the garden of course, which has been developed using recycled paving left over from the original garden.

Life in our ecohouse

The house has proved to be very comfortable, and the children love the space for them to run around in. The project proved more ambitious than I imagined: it has cost more than the back-of-envelope budget because of the extra labour we had to employ, and because we chose to go for the best and not cut corners. It took longer than envisaged largely because of the complexity of the design, which is a far cry from the simple, light constructions that Walter Segal designed. On the other hand it proved much easier than I expected to build: I thought the form of the building would lead to all sorts of problems, whereas in practice timber construction is so adaptable that we were able to overcome all difficulties. There are surprisingly few things I would do differently. I would make the house much more energy-efficient and sustainable if I was doing it now – although we were constrained by cost to some extent. I would pay more attention to ease of maintenance and doing things properly rather than taking a short cut in the heat of the moment just to get the job done – but I was up against time to some extent. I could not sustain the level of effort required to keep going as a director of a company and also as a spare-time builder. In practice the need for maintenance has been very low, but somehow after enjoying the luxury of a new house which needs little or no maintenance I resent the fact that inevitably as the building gets older it will need more attention. I would design the building to be much more adaptable; we now have two teenagers and no au-pair, so the need for accommodation is different from when the house was designed. We do not need a playroom. I work at home now, rather than Rona, and I could do with more space as I now employ an assistant.

Self-build can have unexpected effects on people's lives; our two children try not to make a thing of the house when their friends come round, but they did both find themselves on the television at a young age. It was not a good idea to have a new baby in the middle of the process, but we survived and have now enjoyed many years in our new house.

OVERLEAF Left: Looking down into the family room.
Right: Corner of Shaws Cottages at night.

2.2 An integrated sustainable development

This earth-sheltered development of five houses at Hockerton near Nottingham is heated by the sun. This project is one of the most comprehensive responses to the environmental agenda in the UK, and it addresses the social and economic as well as the environmental dimensions of development. The development is very far-reaching because it has eliminated the need for imported energy and water, has a low environmental impact, and has created local economic and educational activity. The self-builders are also committed to a programme of activity aimed at persuading others to follow their example.

Introduction

This project is an innovative sustainable development in the village of Hockerton near Nottingham. It was completed in September 1998 after three years of planning and 18 months of construction. The houses have been designed as one of the first zero-energy residential systems in the UK reducing life-cycle energy to a minimum. The houses are amongst the most energy-efficient purpose-built dwellings in Europe. Maximum use of benign, organic and recycled materials has been made in the construction, and the development is designed to be to a large extent self-sufficient. The houses are earth-covered, and have passive solar heating without a space-heating system. A wind turbine and photovoltaic system provide all of the energy required to run the homes. The water and sewage system is self-contained.

It is the UK's first earth-sheltered, self-sufficient ecological housing development. The project is noteworthy because it integrates the principles of reducing environmental impacts, the needs and

Above: This community of five households is self-sufficient in energy, water and much of their food. Below: Blinds reduce the risk of overheating in summer.

Hockerton: a coming together of a number of inspired and energetic individuals.

expectations of families and the local community, and the need to create a viable position within the local economy. This is a properly sustainable development in which ecological impacts, social relations and economic necessities have been considered and accounted for.

Hockerton shows how national targets for reductions in energy consumption and emissions can be achieved without compromising modern comfort and amenity. It also shows how small communities can enjoy benefits from sharing some living and working arrangements.

The project was also underpinned from the start with wider aims beyond just creating a congenial, sustainable living environment for five families, aims which were to demonstrate that ordinary families can take responsibility for and contribute to reducing environmental impacts and global warming and to promote this approach to a wider audience.

How did it start?

As is so often the way with good ideas, the Hockerton Housing Project came out of the coming together of a number of inspired and energetic individuals who all brought something to the enterprise. David Pickles was the Chief Architect and Energy Manager for Newark and Sherwood District Council who had, over a number of years, persuaded the council to champion energy conservation and establish an Energy Agency, and to promote the Sherwood Energy Village. The District became an energy-conscious hotspot, and granted planning permission in 1993 for Robert and Brenda Vale, leading green architects of international renown, to build the Autonomous House in Southwell, just down the road from Hockerton. This was the first project to demonstrate how a zero emissions dwelling could work. It also showed how the environmental impacts of the construction of the house itself could be reduced. The Autonomous House was built by Nick Martin, a local builder,

and it was he who, inspired by that example, gathered a number of like-minded people together and commissioned the Vales to design a small group of sustainable dwellings on a parcel of land in the ownership of his family in Hockerton.

Permissions

As often happens with unfamiliar building projects, a number of hurdles had to be overcome – hurdles that often cause projects to fall. The ultimate success of this project is testimony to the enthusiasm and determination of the group and the crucial support of some individuals in the council and elsewhere.

And, as also often happens, the main difficulty was obtaining planning permission. The planning authority was not sure about the idea of underground dwellings and the proposals for a wind turbine. Most importantly, the proposal was for dwellings in the open countryside, which would normally only be granted permission if there was a proven need for agricultural purposes. However, the authority was committed to energy conservation and was familiar with the Autonomous House. The group prepared very detailed reports as part of their submission, and consulted the officers and established a good working relationship before making the application. The proposals to integrate the development with food production and employment opportunities helped, as did the facts that ownership and management were controlled by a legal agreement and that the earth-sheltered dwellings had little visual impact. The development was narrowly approved as it contributed to the council's objectives and policies with respect to energy conservation and renewable energy provision – all subject to what is referred to as a Section 106 Agreement. Hockerton is as a result unique in having a permission to build a sustainable development on agricultural land.

The proposed wind turbine was not approved at this stage, and was subject to a separate application. This was refused on grounds of noise and visual

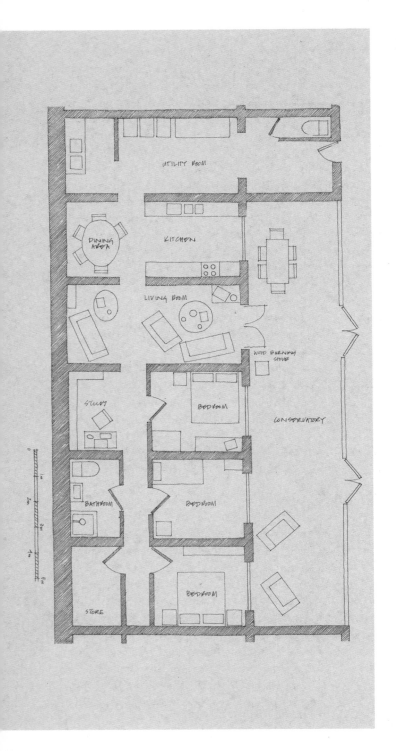

UTILITY ROOM

DINING ROOM

KITCHEN

LIVING ROOM

WOOD BURNING STOVE

STUDY

BEDROOM

CONSERVATORY

BATHROOM

BEDROOM

STORE

BEDROOM

Typical three-bedroom house plan with large conservatory.

intrusion, although the noise was shown to be less than that arising from the nearby main road and although many large electricity pylons are visible from the site – more intrusive, and delivering power from a power station burning fossil fuel and generating pollution in the process! The size of the turbine was reduced and the site moved further away from other dwellings. A second and a third application were refused in the face of concerted local opposition from a small group acting to conserve the countryside. A fourth application was successful, and subsequently permission for a second turbine was granted without hesitation as it became obvious that they are not noisy or dangerous, and look rather elegant poking up above the trees.

The legal set-up

The Section 106 Agreement legally binds the applicants to carry out the development as approved. It refers to a Background Document that enshrines the aims and vision of the project and a Land Management Plan that defines the current state of the land and its biodiversity and how this must be developed in the future. This took two years to agree.

Three legal entities were set up:

- a temporary partnership to manage the construction of the development, which has now been wound up.

- a co-operative company that is responsible for managing and maintaining the project. This issues a lease to the occupiers which includes amongst other things specific conditions, such as a restriction to one fossil-fuelled car for each household and an obligation to contribute a minimum of eight hours per week to maintaining the energy and water systems, cultivating the organic vegetables and fruit, tending bees and fish in the lake and contributing to the outreach activities of the group.

- a co-operative trading company that manages the commercial activities such as consultancy associated with the project.

Finance

Due to the unusual nature of the development and the fact that it was a self-build project, many conventional lenders were unwilling to consider financing the project. However, the Co-operative Bank provided development loans, which were later converted to mortgages with the Ecology Building Society. These two organizations specifically support projects with environmental aspects.

Earth-sheltered building

Traditional human settlements make use of the local topography and microclimate to provide protection from extremes of weather, to reduce the impact of cold winds and excessive heat. People have made use of local materials with minimal environmental impact, well suited to the local climate, to create the heavy mud walls of the deserts of New Mexico and North Africa and the lightweight timber platform houses of the hot damp tropics in Malaysia and the Caribbean. This zero-energy, passive approach to moderating the climate has been replaced with heating, ventilating and air-conditioning systems, all reliant on cheap fossil fuel to create comfortable internal conditions. Nowadays, energy-intensive environmental systems permit designs which are not related to the climate, buildings with large areas of glass for example, which would be unusably hot in summer and too cold in winter without energy-intensive heating and cooling systems.

More recently there has been interest in designing with the climate in mind – a so-called bio-climatic architecture – with the aim of creating more energy-efficient and comfortable buildings by combining traditional forms and materials with appropriate modern technology. Earth-sheltered building is one expression of this. It relies on the very constant temperature of soil moderating the temperature around the building, so that the difference between the temperature inside and outside is half what it would be for a fully exposed free-standing building above ground. The soil acts as a thermal store, and so on a hot day heat is transferred to the soil, creating a cooling effect. At night the temperature drops and some of this stored energy is released back into the house. There is also a thermal lag between soil temperature and air temperature from one season to another, so that soils warmed in summer give off their energy in winter to warm the house and in summer the soil is cooler than the air and will absorb heat to provide cooling.

Earth is a cheap, natural, local material and can reduce energy consumption and emissions. It will provide an airtight building and protect it from the weather and reduce maintenance. It can create roof gardens and create new habitats which can be particularly important in urban areas and reduce the visual impact of a building.

The Hockerton building

The Hockerton Housing Project consists of a terrace of five single-storey dwellings which are earth-sheltered at the rear (north) such that the ground surface slopes and blends smoothly into the field at the back. Each house is 6m deep with a 19m south-facing conservatory running the full width of each dwelling. A repeated modular bay system of 3.2m in width was used for ease of construction. Four of the houses are of six bays and have 120m² net internal floor area and 47m² of conservatory. Most of the internal rooms have 3m-high French windows that open to the conservatory. Those rooms that are not so dependent on natural light, such as utility and bathing areas, are located towards the rear of the homes. The entrance to the houses has a lobby which acts as a buffer between the inside and outside to minimize the escape of warm air.

The site

The development is located on a 10ha site that has a slight slope just to the west of south. Previous use of the land was essentially agricultural. The large area has allowed the incorporation of features that enable the occupants to live in a sustainable and self-sufficient way. The group produce around two-thirds of their requirements for vegetables as well as some fruit, eggs and meat. This makes a significant reduction in the need to bring in food, which for a typical family now results in twice as much carbon emissions from fuel, chemicals and transport as central heating or running a car.

It has also allowed for large water catchment for the homes and waste disposal via a reedbed system, and 4000 trees which absorb CO_2 as they grow have been planted to offset the carbon emissions arising from the use of large quantities of Portland cement in the construction and the use of mechanical plant during the building process.

The group also run an electric car and use bikes and share cars to reduce their use of fossil fuel for transport by around 50% – in a village with no public transport. This raises the question of density. It is argued that it is more sustainable to build at relatively high densities because this reduces the need to travel to work, to the shops and to school, and because it also makes public transport viable. However, it should be noted that cities still rely on a huge hinterland to supply them with food, water and space to dispose of waste.

Design objectives

The key design objectives regarding energy performance and sustainability were:

- reduction to zero of the need for space-heating by artificial means
- reduction to zero of CO_2 emissions incurred by the existence of the development

Above: The houses are heated by the sun.
Below: The rooms are bright and light,
in spite of being underground.

- to be as autonomous as possible in terms of provision of utilities, including water

- to use renewable energy sources to meet the energy requirements of the development

- the use of easily transferable construction techniques and readily available, environmentally responsible materials

- competitive costing to conventional housing in the short term, with demonstrable savings over conventional housing in the medium to long term

- occupier control of infrastructure and services with minimal maintenance

- increased biodiversity and enhanced landscape associated with the project

- offsetting all carbon emissions (including those embodied within the materials) and CO_2 emissions incurred during construction work

- to achieve all of the above with no loss of comfort or modern amenities

Building construction

The development is of high thermal mass construction, having 200mm concrete block internal cross walls on a 300mm concrete slab, a concrete beam-and-block roof and 500mm-thick external walls of two skins of concrete blockwork used as formwork to contain mass concrete. A polyethylene waterproof geomembrane waterproofs the building from the surrounding soil.

Walls, slab and roof are super-insulated with 300mm of expanded polystyrene (CFC-free) with the mass on the inside of the insulation. The roof is covered with 400mm of topsoil, and the north side and ends of the terrace have soil heaped up and over them. The building envelope is clay brick for the exposed exterior walls, using bricks fired from

waste methane gas. All of the internal walls are wet-plastered. There are no holes through the main slab for soil pipes or services, so the insulation and membranes are not perforated.

The main doors and windows opening into the conservatory are triple-glazed with low-e glass and argon filling, whilst the conservatory has double low-e glazing. The solar space-heating is completely passive; heat transfer from the conservatory to the house can be facilitated by opening the windows if required.

The roof, walls and floor have a U-value of 0.11 W/m^2K and the triple-glazed units 1.1.

The underground concrete construction is a very airtight way of building. The test results show that the mean air permeability of the dwellings at Hockerton (excluding the conservatory) lay between 0.95 and 1.23 $m^3/h/m^2$. (This measure is the number of cubic metres of air that passes through a square metre of exposed construction – wall, roof or suspended floor – in one hour, all at a standard pressure difference between the inside and outside of the building of 50 Pascals. This pressure difference is higher than would normally be experienced even in windy conditions, and so the actual amount of air leakage is generally substantially less than the test result.) This result is considerably lower than the UK mean of 11.48 $m^3/h/m^2$, and the maximum specified level of 10 $m^3/h/m^2$ in the Building Regulations.

When choosing materials, emphasis was placed on high performance, low maintenance and long-term reliability. Readily available, low or medium tech local materials were used where possible to reduce the energy embodied in the material from manufacture and transport. Potentially toxic raw materials and finished products were avoided to minimize risks to the environment and to health. The environmental impacts on the land and resources as well as the environmental policies of the manufacturers were considered.

The use of PVC was avoided as it is potentially carcinogenic and gives off toxic gases in a fire.

Wiring which does not use PVC insulation was used, and copper gutters, clay drainpipes and clay floor tiles were used instead of PVC products. Boards that give off formaldehyde gas from the resin in their composition were avoided where possible and paints low in Volatile Organic Solvents (VOCs) were used.

Ventilation system

Ventilation is provided by opening windows in the external wall and glazed roof of the conservatory, and opening windows and glazed doors between the house and the conservatory. In addition, each house has a mechanical ventilation heat recovery (mvhr) system that supplies fresh air to the living rooms and bedrooms and extracts from the kitchen and bathroom areas. This ingeniously uses large diameter clay drainpipes suspended on chains as ducts.

Energy-saving appliances

Originally each house was supplied with hot water that had been heated using an air-to-water heat pump. The system is maximized by drawing air from the top of the conservatory to gain the benefit of solar heating and stratification of the conservatory air. Heated water is stored in a heavily insulated 1,500-litre plastic tank in the utility room/laundry of the house. This provides hot water using less than a third of the energy required for a conventional system. However, these units have proved unreliable and the manufacturer no longer supports this equipment. All but one of the houses now obtain hot water from an immersion heater running on the renewable power generated on site.

Other energy-conservation measures include predominant use of low-energy light bulbs, laptop computers, and appliances that are highly energy-efficient. Appliances are not left on standby, and clothes are dried on conservatory racks rather than in tumble-dryers.

Solar energy utilization

Space-heating relies totally on heat from solar gain and incidental gains from occupation. The heat is stored in the mass of the buildings (e.g. concrete and blockwork) and released when the air temperature drops below that of the building fabric.

The design is enhanced by the roof angling upwards which makes good use of low winter sun penetrating to the back of the dwellings. This design provides good internal daylighting as well as maximizing passive solar gain through the conservatories. The trees on the southern boundary are all deciduous, reducing blocking sunlight once they lose their leaves in autumn. During the summer, shading is created within the homes due to the high angle of the sun – this reduces thermal gain and brightness inside when it is least wanted. The houses are earth-sheltered, with a minimum of 400mm of soil on the roof, which provides protection from the dominant cold north-easterly winds during winter.

Renewable energy systems

A 5kW wind turbine and a 7.65kW array of photovoltaics generate almost as much energy as is used by the homes. The HHP wind turbine is one of very few examples in the UK of a community-owned wind turbine, whereby the owners are supplied directly with the 'clean' renewable energy produced. The photovoltaic system has been part-funded through a DTI photovoltaic domestic field-trial grant.

The wind turbine is expected to produce around 12,000kWh annually, given the wind conditions on site, with the roof-mounted photovoltaics producing a further 6,000kWh. Both systems are grid-linked, which allows for both the import of energy during periods of supply shortfall and export during periods of excess energy production. The excess exported will offset the imported energy from the grid.

Left: Turbines at sunrise. Right: Photovoltaic panels together with the windmills generate as much energy as the development uses.

Water collection and treatment

The high-grade drinking water requirements are met by the collection of rainwater from the conservatories. The water is stored, filtered, mineralised and pumped to drinking water taps in the houses.

The low-grade or non-potable water is collected from the back of the houses, road and surrounding fields. It is then channelled into a 5m³ underground tank, from which it is pumped to a reservoir. The water goes through a sand filter before reaching the houses. This removes particles and has some bacteriological action. The reservoir store of 150m³ of water is equivalent to approximately 100 days use by the project members.

Sewage and wastewater are initially treated in a septic tank with a 10-day retention, after which the outflow runs to a floating reedbed. The arrangement of the reedbed ensures a very long dwell time

in the reeds – probably three months – to ensure complete purification before it passes through a limestone gabion wall and into the main lake.

Energy performance

A monitoring programme (New Practice Profile 119 published by the Energy Efficiency Best Practice Programme) was conducted over the first year of occupation only (1998/1999). The total energy consumption of the five homes during the monitoring period was just over 4,000 kWh per house, and around 11kWh per house per day. This equates to an energy use of about 25% of conventional new UK housing, and only about 10% of the energy use of the average of the current UK building stock including existing dwellings. Recent monitoring shows energy consumption for the different houses varies between 16 and 24 kWhr/m²/yr. Monitoring of internal temperatures shows that the minimum temperature in winter is around 18°C.

The temperature variations between the houses are dependent on occupancy rates and desired comfort levels. The houses are all electric, so any use of heating appliances will be incorporated in the figures given. The energy consumption of some of the houses is higher than the initial estimated figure of 3,000kWh/pa. The major factors affecting energy consumption are home-working and teenagers. The occupants may live in energy-efficient houses, but lights and stereos still get left on and doors still get left open!

Each of the houses has a wood-stove in the conservatory. These are used occasionally during the winter months, mostly for social events, but also provide a back-up for cooking in the event of power cuts. The stoves contribute nothing to the temperatures in the houses other than to marginally reduce heat loss in localized areas of the front façade of the main house for short periods of time.

Costs

The Building Research Establishment (BRE) compared the average construction cost of a traditional four-bedroom detached house of 125m² gross at £425/m², with the cost of a Hockerton house of the same size at £485/m² (all at 1997/98 prices).

However, these figures exclude the HHP conservatories, which are arguably, used more by the occupants than other parts of the house. The costs for the HHP conservatories were approximately £255/m², which if included in the total floor space would decrease the overall cost to £421/m² gross – about the same as the conventional detached house. It must be remembered however, that members of the group provided around half the labour to build the scheme. This self-build labour reduces the cost by around 20% of the total cost of the scheme if it was built by a contractor in the conventional manner.

It must also be remembered that running costs for energy and water are saving the occupants around £1,000 per year – year after year after year.

Although there is a large amount of concrete and insulation in the Hockerton design, the form of the houses is relatively compact, the construction is simple, and the finishes and fittings in bathrooms and kitchens are economical. The emphasis is on reducing running costs and emissions into the future.

Spreading the word

The group have been keen to encourage others to follow their lead and create other sustainable developments around the country. They have published excellent information on the Hockerton project, and have maintained a busy schedule of visits and courses for people to find out about their experience. They also run events in partnership with local organizations and schools, so that the project is seen more as an integral part of the local community and not a group of slightly odd eccentrics with a bee in their bonnet. Nick Martin has gone on to build a pair of similar earth-sheltered dwellings on land adjacent to the site, and now has a flourishing business constructing sustainable buildings.

To meet the growing requirements of visitors, a new learning resource centre has recently been completed, which includes a dedicated audio-visual room, seminar facilities and a permanent exhibition. It is the base for a new business providing consultancy and training in sustainable development, and is seen as a regional catalyst for sustainable action of all kinds. The grass-roofed, single-storey building has been built to similar high environmental standards as the homes, and is designed to demonstrate the features of sustainable building practice.

2.3 A group of tenants build in the inner city

Self-build is often assumed to be low-density rural or suburban development, but this project in Islington in North London shows how it can work in a dense inner city neighbourhood. The development was completed in 1995, when standards of energy-efficiency and environmental performance were not as high as currently – which just goes to show how fast things are changing at the moment. However, the scheme is well insulated with cellulose fibre made from recycled newspaper within a timber frame, which is an inherently very sustainable form of construction. The scheme is *built using the Segal method of post and beam timber-frame construction devised by the architect Walter Segal in the mid-1960s and subsequently used with great success by many self-build groups. This method is very adaptable, and the members of the self-build group were closely involved in developing the design and building the houses. The buildings can be adapted to changing needs and expectations in the future, which is a necessary feature of any sustainable system. So too is ensuring that residents have an active role in the process so that they understand how the houses work and know how to use and maintain them effectively.*

In the 1970s, with the encouragement of the then Labour government, Islington Council in North

Two- and three-storey houses.

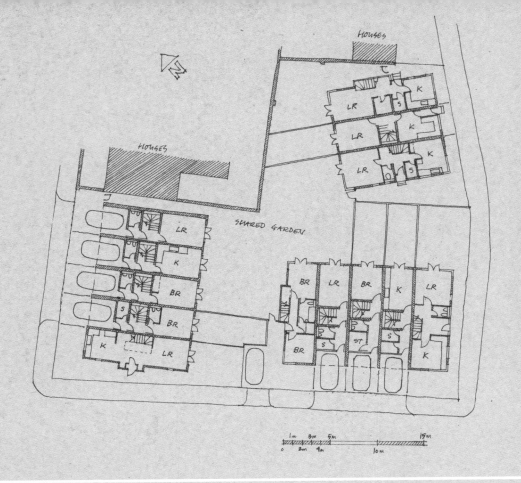

HOUSES

HOUSES

SHARED GARDEN

LR   S   K
LR   K
LR   S   K

LR
K
BR
S
BR
K   LR

BR   LR   BR   K   LR

BR   S   ST   S   K

1m 3m 5m   15m
0   2m   4m   10m

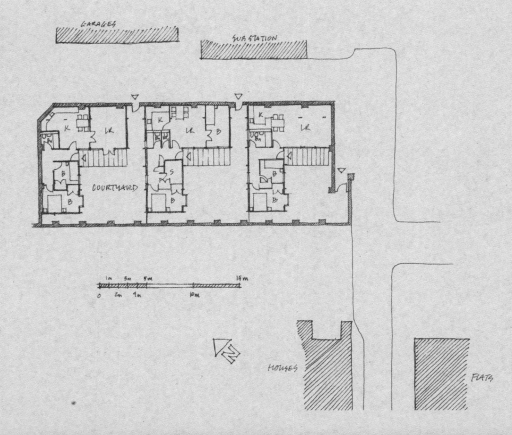

GARAGES

SUB STATION

K   LK
B
B

K   LK   B
S
B

K   LK
B
B

COURTYARD

1m 3m 5m   15m
0   2m   4m   10m

HOUSES

FLATS

London bought up many decaying properties with a view to refurbishing them and letting them to needy households at rents they could afford. Due to a lack of financial resources the council resolved to let many of these properties to members of a short-life housing co-operative at peppercorn rents rather than see them remain empty, but when later faced with a critical shortage of affordable accommodation for needy families, the council wanted to take these properties back. However, they could not throw many of the present occupants out on the streets, and so the idea of a self-build scheme as one form of move-on accommodation for these short-stay tenants took shape. Many of these people had been carrying out self-help maintenance and improvements to their dilapidated short-life properties for years, and so it made sense for them to go on and build their own houses. A small housing association was established, a self-build group was recruited, and two council-owned sites were identified.

Finance

At this point one of the apparently frequent delays occurred, and they came from unexpected directions. Firstly, Prince Charles took a drive around the East End of London under cover of darkness, and was appalled by what he saw. It was suggested that people building their own homes could help provide good homes at affordable cost. The chief executive of a prominent property company put up a (large) sum of money to develop the legal framework and documents to establish a financial model to fund self-build developments by getting private landowners to make sites available on a deferred payment basis. It was a hugely complicated arrangement, but was attractive to the council because it relied on private money and did not require any public money, which is always in short supply. In the end this initiative failed, and only one or two developments were completed using this model.

So it was back to square one on the funding – and

Some of the self-builders chose to combine their private back gardens into a shared garden.

on to the Housing Corporation, who by this time had been persuaded to make funding available for what might be described as 'social' self-build schemes. The first social self-build had been sponsored by a local authority, but the by now Conservative government had stopped local authorities funding housing. The Corporation funding was for shared ownership, where the self-builder owns part of the equity or value of the house, for which he or she takes out an individual mortgage, and the remainder of the equity is owned by the housing association, for which they pay a proportion of the normal rent for the property. The equity owned by the self-builder is generally wholly or largely paid for by the 'sweat equity', that is the difference between the cost of building the house using unpaid self-help labour and its value on completion. In a high-value area such as Islington, the sweat equity can be substantial, whereas it can be dif-

FACING PAGE Above: 13 narrow frontage two- and three- storey town houses with a shared garden.
Below: 3 single storey courtyard houses on the site of an unused play area on a council estate.

ficult to make this arrangement work financially in low-value areas such as the north of England. Shared ownership reduces the cost of living in the house at the outset whilst allowing the self-builder to purchase a greater proportion of the equity, including outright purchase, at some time in the future. It is a very flexible form of tenure. Unfortunately, around this time the original housing association was wound up, and time was lost getting a larger, more established association to take over the scheme, and most importantly, administer the grant from the Housing Corporation.

The sites

Meanwhile, the self-builders had been working with the architects to develop the design. This was for 13 narrow-fronted two- and three-storey town houses for one of the sites, where they took their place in the street with Victorian two- and three-storey terraced houses, some of which were beyond the end of their useful life and had been demolished to make way for the new development. The other site was again not untypical of urban areas – this time an unused area within a council estate. It had originally been designated as the mandatory play area required for a development of council flats. It is surprising how many small redundant sites exist within urban areas. One local authority recently identified space for over 400 houses on such sites in their area, which they are now developing with new houses.

The Islington play area was a tarmac yard surrounded by a 2m-high wall – and nothing else except a drain grating in one corner. It's astonishing that so little care and thought could go into creating an environment for living. The yard was not overlooked, and so the local youth could get up to all sorts of mischief and residents could dump unwanted junk, all without fear of discovery. The proposal was to build three bungalows, each looking into their own small courtyard garden within the surrounding perimeter wall.

The Segal Method

Three basic house types were designed: a two-storey town house with two bedrooms, a three-storey town house with three bedrooms, and courtyard bungalows with two bedrooms. The buildings were all timber-framed and based on the so-called Segal Method of post and beam construction. This form of construction was devised by the architect Walter Segal in the mid 1960s, and is based on the idea of combining readily available, standard building materials (originally including wood-wool slabs and plasterboard) within a post and beam timber frame using dry fixing methods such as bolts and screws. The building is laid out on a modular grid – based on the stock sizes of the materials – to avoid cutting and waste. The method is devised to be economical, and quick and relatively straightforward to construct.

Adaptability

The external and internal walls are not loadbearing, as the timber frame carries the weight of the building. This means that one can place doors and windows in any position on the outside of the building and can position walls anywhere you like on any floor on the inside. Importantly it also means that you can change your mind as you go along. You can stand in the part-completed building, see which way has the view and which direction the sun is coming from, and adapt the layout to suit. This can be very useful for self-builders, who often find visualizing the spaces inside and outside of buildings difficult from the drawings used by architects and builders to communicate the form of buildings. The separation of frame and infill means that the houses can be relatively easily adapted to changing needs and expectations in the future. This flexibility is a necessary part of any sustainable approach to house design in my view, as it ensures a long useful life for buildings, yet flexibility is not generally considered important in house design in Britain.

Single-storey courtyard houses create a quiet haven within the inner city.

Participation

The self-builders embraced the idea of freedom of layout and openings with enthusiasm, and soon no two houses were the same. The self-builders also had the choice of materials, fittings and colours inside. In my view, this ability to be involved in the design of one's own home is a necessary element in a sustainable approach to housing. The role of people in sustainable development was stressed at the Rio summit in 1992, yet, like flexibility, participation remains tokenistic generally in thinking about housing in the UK. The process of designing the houses kept up the self-builders'

commitment through the planning stages whilst the finances were put in place, and during the hard slog of building on site.

The layouts of the two sites demonstrate contrasting approaches to high-density development. One was of tall narrow houses with back gardens, many of which the self-builders decided to combine into one larger communal garden, with each house having a small private patio immediately next to the house. The three-storey houses also had a balcony at the back opening off the first-floor living-room. The other site was developed with low L-shaped houses planned around a small but completely private courtyard.

> *This development puts land to good use which had previously been unused, and it shows what can be achieved on those small sites which exist in our cities.*

The party walls

The construction incorporated a high level of insulation to walls, ground floors and roofs provided by cellulose fibre insulation made from recycled paper. The two- and three-storey terraced houses have concrete block party walls intended to provide good sound and fire separation between the houses. Special blocks were used, designed to be laid dry without mortar, to make block-laying a job that could be undertaken by the self-builders rather than having to employ skilled masons. Steel reinforcing rods were placed through the hollow cores of the blocks, which were then filled with concrete to make a strong and solid wall. Unfortunately this form of construction did not prove to have as good sound insulation as was hoped, and one or two self-builders complained that they could hear what was going on next door. It seems that the walls were acting as a large drum, transmitting structure-borne sound to the adjoining house. I have worked on a great many timber-frame houses, and not once have I experienced a problem with sound transmission through a party wall – which is not what most people would expect.

The cladding

The buildings are largely clad in green oak (oak that has not been seasoned – either naturally under cover for a year, or in a kiln in a matter of a day or so). It is a relatively plentiful and cheap, durable, and does not require an applied finish and is therefore maintenance-free. It weathers on the building to a grey colour, but does have a tendency to twist and warp as it dries out. This lends a certain rustic quality to the building, but you may find that you have to replace one or two boards that are too twisted to stay in place.

Trainees

Most of the self-builders were working throughout the construction period, but they did employ a site manager who was able to supervise a group of carpentry trainees from a local training centre. The trainees installed staircases and proved to be a useful extra resource. For their part, opportunities to work on-site are limited, and they obtained site experience in a supportive environment.

The potential of the inner city

This development puts land to good use which had previously been unused and it shows what can be achieved on those small sites which exist in our cities. It also demonstrates how self-build houses can be built at the higher densities necessary to be cost-effective on expensive land in the heart of big cities. The development provides affordable homes of a high standard, which can be a route for young people onto the housing ladder in a very high-cost area.

2.4 A three-bedroom eco-house for £30,000

This house in Basildon in Essex demonstrates how self-build can create fantastic value for money – a low-energy, three-bedroom, two-bathroom family house built by its owner in 1996. The house shows another way of building in timber: this time with load-bearing panels for simplicity and economy, but on timber floor beams spanning between concrete pad foundations which significantly reduces the extent and cost of the foundations.

A plotlands plot

John Little and his now wife Fiona Crummay had restored a 1930s semi, but were now looking for fresh challenges. They had an interest in wildlife, and saw the potential of a near derelict four-acre plot in Laindon near Basildon, not far from where they had been living. The plot came complete with a tiny little two-room house built in the 1920s. The site is in an area of small plots which were made available at low cost during a period of agricultural depression. Many working-class families from the East End of London were able to buy plots and develop them at the outset with a small weekend cabin. Many people subsequently moved out of the city and established new communities scattered around Kent and Essex, the east and south coasts and in the Thames Valley, as Colin Ward and Dennis Hardy described in the preface to their book

Arcadia for All. The site is around the corner from the Plotlands Museum, where an original plotlands house has been preserved.

In the first half of the twentieth century a unique landscape emerged along the coast, on the riverside, and in the countryside, which was more reminiscent of a frontier than of a traditionally well-ordered English landscape. It was a makeshift world of shacks and shanties, scattered unevenly in plots of ranging size and shape, with unmade roads and little in the way of services.

To the local authorities it was something of a nightmare – an anarchic rural slum, always one step ahead of evolving environmental controls.

But to the plotlanders themselves, Arcadia was born. In a converted bus or railway carriage perhaps, and at a cost of only a few pounds, ordinary city dwellers discovered not only fresh air and tranquillity but, most prized of all, a sense of freedom.

A modern plotlands house

John and Fiona's plotlands house was very charming with its little veranda, but the bedroom was not much larger than the bed and they were keen to start a family. The house had almost no insulation and was freezing cold in winter. It had an asbestos cement roof that was beginning to crumble, and its timber frame was beginning to rot. It would have to be replaced, but John and Fiona were keen to build in sympathy with the plotlands idea. The new house would be built of timber to be economical, human in scale, simple and unassuming, but provide modern standards of

Left: The house is sunny and bright and faces south towards the view. Right: In an area of plotlands development, this timber bungalow is designed to be sympathetic to its simple and economical neighbours.

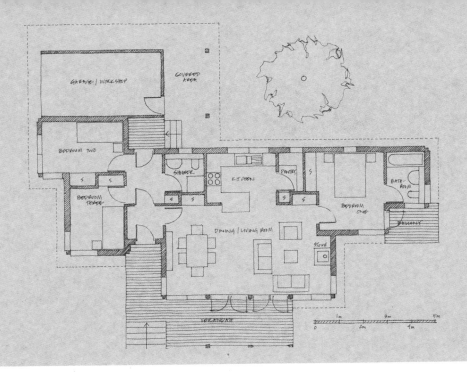

Above: The three-bedroom, two-bathroom plan faces south.

comfort. With the passing of Town and Country Planning legislation starting in 1948, the area became designated as green-belt land, to be protected from development. However, the planning authority was happy to support the idea of a modest, single-storey, weatherboarded building designed to be sympathetic to the plotlands model to replace the existing house. They agreed to relocate the new building to a position higher up the slope of the site, which meant that the house enjoyed a panoramic view to the south, with the Thames and north Kent hills in the distance. The house was furthermore sheltered from the north-east by trees and a high, dense hedge.

The layout of the house

The simple, single storey plan is laid out with rooms arranged along the contours, all facing south, starting with the principal bedroom and en-suite bathroom at the east end. This room has its own small private balcony. Then comes the large living-room, with space for a dining-table, and a wood-burning stove with the kitchen and walk-in larder in an alcove at the back of the room. There is an extensive deck that is partly roofed with white panels, which protect

people and building from the weather, provide shade both inside and out, and serve to reflect light through high-level windows up onto the ceiling and back into the living-room. There are two small children's bedrooms and a shower room at the west end. The building is tucked into the hill under a grass roof made from turf cut from surrounding meadow. The roof slopes down over the garage on the north side, reducing exposure and bringing the grass roof down almost to meet the meadow behind the building. John has since built a number of outbuildings and sheds, all with grass roofs, which look wonderful and make the buildings fit together visually and look as if they have always been set within the landscape.

Green roofs

Green roofs (rather than grass roofs, for they can support a variety of plant life, not just grass) are not just good to look at, for they have a number of practical benefits – they create new habitats on the roof to replace those lost by building on the ground, and these habitats support a wide range of wildlife, including rare insects, as shown by a British Nature study of green roofs in urban areas. A green roof also holds a large amount of water after it rains, and thereby reduces storm-water runoff. This reduces the load on the storm-water system and reduces the risk of flooding. This is a particular issue in urban areas, and London in particular, which has a combined drainage system. The sewage treatment works have a limited capacity, and so periodically when there is heavy rainfall, large quantities of untreated sewage are dumped in the Thames. The soil of a green roof acts as a thermal buffer and will tend to keep the building warmer in winter and cooler in summer.

Contrary to popular mythology, it is not necessary to mow or maintain a grass roof; it can be left to grow in its natural state and attain a balance of different plants which will grow in the conditions. The shallow depth of soil will limit growth, and the vegetation will go brown and die back during dry spells in summer. It is possible to incorporate measures to retain moisture in the roof, but this can be very expensive and is never very effective. It is also possible to create more of a garden effect by having a substantial depth of soil and planting significant plants and shrubs, but it is expensive supporting the weight of soil.

It is possible to use a lightweight system consisting of a plastic mesh containing a seeded lightweight growing medium. This reduces the cost of the supporting roof structure, but is itself expensive. John and Fiona's grass roof on the other hand is simply turf from the paddock at the back of the house laid on around 75mm of soil on a layer of geotextile membrane which protects the single-ply roof membrane. The roof is pitched at about 15^0, and drains to a shingle margin laid behind an upstand at the edge of the roof which retains the soil on the roof. A green roof can be at any pitch between about 30^0 and dead flat, but at a pitch of 30^0 or so, netting or battens will have to be incorporated into the construction to prevent the soil from sliding. It is important that the soil is of poor fertility to encourage wild flowers which can survive the hot, dry summer conditions and to discourage those species which cannot.

The structural arrangement

The building is conceived as a conventional timber-frame structure with wall panels, but instead of being supported on a continuous foundation below the walls, it is supported on concrete pillars. Timber beams span between these concrete posts at ground-floor level. This arrangement combines the advantages of the simplicity and economy of timber-frame panel construction with the benefits of the simple and economic foundations of a Walter Segal method post and beam frame. A conventional stud panel construction made of timber studs nailed together at 600mm centres is easy and cheap to build, but generally requires to be fully supported on a continuous foundation. A Segal house, on the other hand, is supported on calculated timber columns bolted together between 3 and 4 metres apart, each standing on a separate isolated concrete pad foundation 600mm square and 900mm deep. The structural frame has to be made of high quality timber, and is generally more expensive than a stud frame, but the extent and cost of the foundations is substantially less. The combination of panel construction supported on isolated pad foundations makes the most of both systems by combining an economical structure on a reduced foundation arrangement.

Cost-effective foundations

I estimate that relying on isolated bases below each post can reduce the extent of the foundations to around 10% of the foundation needed to support a conventional brick wall. It also removes the necessity of levelling the site, and so John and Fiona's house appears to float above the landscape. Steel reinforcing rods had been cast into the concrete bases and left sticking up. Drainpipes were used as permanent shuttering and placed over the reinforcing rods which were then filled with concrete to form the supporting columns.

Energy-saving measures

The roof has 250mm of insulation, the floor 200mm and the walls 150mm. This gives U-values of 0.13, 0.18 and 0.20 $W/m^2\,{}^0C$ respectively. The windows are of high-quality double-glazed timber, made in Denmark. The house has a large amount of south-facing windows and benefits from passive solar gains. It has a central heating system installed, but this is seldom used as the wood-burning stove fuelled by wood waste from a friend's arboricultural business based on the site generally provides enough warmth.

Value for money

The house has an area of $100m^2$ and the basic house cost £28,000 to build in 1995. This cost was increased by spending a bit more on some of the equipment such as stainless-steel electrical accessories, elegant radiators and a gas-fired Aga in the kitchen which runs the heating and hot water. This increased the cost to £38,000, but even so this represents stunningly good value for money. Costs were kept low as almost all the work was carried out on a self-build basis. John was taking time off working in the family-run shoe shop, and his brother – a local small builder – and his father who had part-built his own house provided the rest of the effort.

The Grass Roof Company

The house is now surrounded by a lush native wild garden planted with thousands of trees and shrubs and a wildlife pond. There are a number of timber, grass-roofed outbuildings designed and built by John, who gave up the shoe business years ago and has now established the Grass Roof Company with his brother designing and building extensions and small buildings for schools and the like.

The house is set within a wildlife garden.

2.5 Super-economical super-insulation with straw bales

This house in rural Herefordshire is an example of a thoroughly practical approach to designing and building to reduce environmental impacts. It is constructed with straw bales within a timber frame to provide insulation in the external walls. It is an interesting example of building within one's means without the need for borrowing, with development taking place slowly as money becomes available. The builder and owner is an expert in water conservation and alternative sewerage treatment systems, which he has incorporated into his house.

Incremental development

Ten years ago Nick and Sheila bought a bare field on top of a low hill deep in the countryside. They set up on-site in a small caravan acquired for £400. They were both working from the caravan, as well as living in it. Nick recounts how the fax machine was mounted at high level, and occasionally spewed out paper over the person sitting and eating below. They set about taming the landscape by planting a garden and trees, and making it more productive by growing fruit and veg and establishing a coppice of willow and hazel for basket-making.

The next stage of development was to construct an interim house, larger than the caravan, better insulated and more comfortable – but still very compact, with 30 square metres of floor space, half of which was devoted to badly needed workshop space. Income from the next job contributed to the very modest cost of £3,000. The house had two rooms and a compost toilet.

Autonomous living

Nick and Sheila were living off the grid, relying on second-hand photovoltaic panels and a small wind generator linked to a bank of second-hand batteries for the storage of power. In these circumstances you become acutely aware of how much power is consumed by every light and appliance you own. Nick can tell you the power consumption of a television on stand-by, or a printer which is apparently turned off. Although their new house is supplied with power from the grid, Nick and Sheila have a low-power-consumption lifestyle to this day.

The permanent house

After living on the site for five years, sufficient money had been accumulated to embark on the permanent house. The initial budget was £40,000 for a 100 square metre house – although that area does not include the porch and two mezzanine areas which make good use of the space under the lofty

Timber cladding conceals the straw-bale insulation.

roof and bring the size of the house up to 130 square metres. The cost rose to around £50,000 over the four years it took to build, fitting the construction around the demands of work and the need to generate cash for building. The ground floor slab was constructed two years before the timber frame was erected on it, the roof followed a year later, and this sheltered the straw before the external walls were completed. It then took another year to complete the house to the point of moving in.

The permanent house is a straightforward, practical design, with every aspect of design and construction carefully thought out and beautifully built. The final result is a house with a very special sense of quality and very good performance.

The layout of the house

The layout developed over a number of months, during which time Nick and Sheila got to know the site and tried out different arrangements in model form. In early designs, the house was long and thin with the rooms facing south with extensive windows. The house they built is much more compact, with relatively few windows facing the view and the garden to the west; instead the house turns its back on the winter storms that come from that direction. It consists of a large main room with a kitchen at one end, a sitting area around a wood-burning stove at the other and an area for eating in between. There is a bedroom, bathroom and toilet off to one side, with a mezzanine above and a home office and visitor's bedroom with a second study mezzanine above off to the other side of the main space. There is a large porch at the front which keeps the heat in and the winter weather out, and which has a sink and the freezer, and serves as a utility room. This is a brave and unusual idea to enter the house through the place usually reserved for keeping things out of sight around the back somewhere. A conservatory is planned, to lead out to the garden at the back.

A timeless interior

The interior draws inspiration from Christopher Alexander's 'Timeless Way', and has a number of interesting corners and places to sit, with deep window seats and an alcove with a small settle. The ceiling is high, with a rooflight, and is supported on high posts with knee braces with traditional pegged joints. The stair to the mezzanine rises on one side and turns through a right angle avoiding the structure and forming the alcove below. This rather picturesque arrangement was the product of a fortunate mistake: as can happen, the stairs needed to occupy a larger space than originally envisaged, and so a change of design was required during the course of construction.

Although the house is timber-framed, it has a very solid feel to it which derives in part from the quality and type of finishes used: the walls and ceilings are wet-plastered, with a slightly rough finish with rounded corners, painted with slightly textured mineral paint in beautiful shades of grey and ochre and purple. The floor is concrete, and was given a fine finish when it was cast with a 'power-float' machine which smooths the wet material with a circular motion. This was then ground to a polish and waxed on completion to give a richly coloured and patterned floor. The stair and fireplace are unfinished natural concrete. The balustrade to the stair of galvanized steel cable and painted steel tube was adapted by Nick, using his welding skills, from a balustrade that was being thrown out of a museum that was being renovated. The bathroom is tiled with travertine on the floor and walls. The generally industrial light fittings were largely obtained second-hand from car-boot sales for around £2-£5 each, and were carefully chosen for their design. The beautiful doors and other joinery were made from Welsh oak by a local joiner. In spite of, or perhaps because of, the simplicity and economy of the materials, the whole effect is one of quality and solidity.

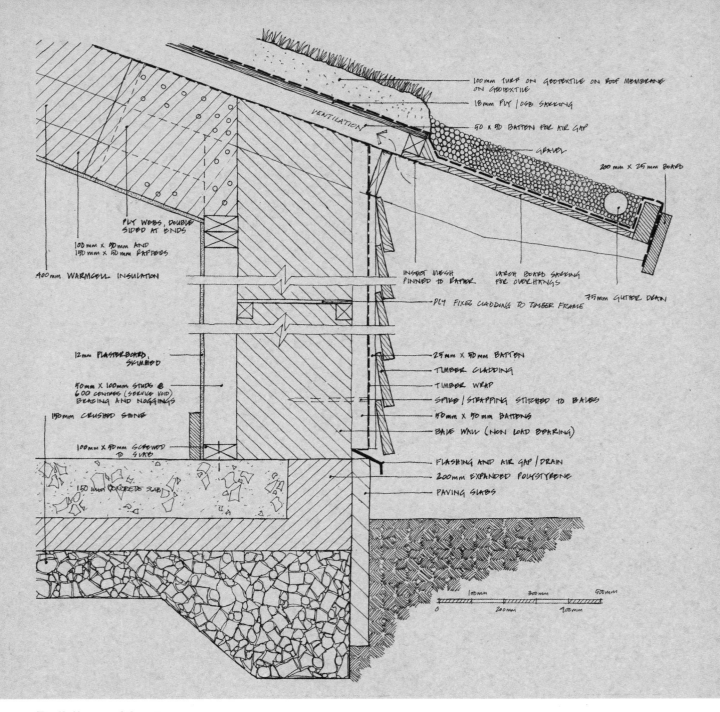

Labels (clockwise/by position):

- 100mm TURF ON GEOTEXTILE ON ROOF MEMBRANE ON GEOTEXTILE
- 18mm PLY / OSB SKINNING
- 50 x 50 BATTEN FOR AIR GAP
- GRAVEL
- 200 mm X 25 mm BOARD
- VENTILATION
- PLY WEBS, DOUBLE SIDED AT ENDS
- 100 mm X 50 mm AND 150 mm X 50 mm RAFTERS
- 400mm WARMCELL INSULATION
- INSECT MESH PINNED TO RAFTER
- WROT BOARD SKINNING FOR OVERHANGS
- PLY FIXES CLADDING TO TIMBER FRAME
- 75mm GUTTER DRAIN
- 12mm PLASTERBOARD, SKIMMED
- 50mm X 100mm STUDS @ 600 CENTRES (SERVICE VOID) BRACING AND NOGGINGS
- 150mm CRUSHED STONE
- 100mm X 50mm SCREWED TO SUKO
- 150 mm CONCRETE SLAB
- 25mm X 50 mm BATTEN
- TIMBER CLADDING
- TIMBER WRAP
- SPIKE / STRAPPING STITCHED TO BALES
- 50mm X 50mm BATTENS
- BALE WALL (NON LOAD BEARING)
- FLASHING AND AIR GAP / DRAIN
- 200mm EXPANDED POLYSTYRENE
- PAVING SLABS

Scale bar: 0 — 100mm — 200mm — 300mm — 400mm — 500mm

Building with straw

The house is essentially a timber stud-framed house with a layer of straw bales on the outside which provides the thermal insulation to the walls. Whilst straw does not insulate quite as well as mineral wool or fibreglass, a layer of bales laid on edge (360mm thick in this case) provides a superinsulated wall at very low cost. The walls of this house have a U-value of around 0.10 to 0.15, depending on the density and orientation of the straw, which was baled on a local farm during a period of fine weather. It is vitally important that the straw is

Above: Detailed section shows the thick straw insulation in the wall protected by rainscreen timber cladding.

baled dry and kept dry, and not discoloured by rot or mould growth. You need a good foundation to protect against rising damp and a deep roof overhang; and in this house there is a rainscreen cladding of timber weatherboarding with a weatherproof membrane and ventilated cavity behind which protects the straw from driving rain.

Straw is cheap, and it cost only a few hundred pounds for all the wall insulation for the house; it is a virtually zero-embodied-energy material, which will compost back into the earth at the end of the life of the building – but you are dependent on the seasons to obtain it, and could not plan it into a tight building programme. Straw bales are dense and do not support combustion, and are also too dense for rodents to become established. One of the effects of straw-bale construction is to create very thick walls, which create deep window sills, good for sitting in and for plants, but which do tend to isolate the inside from the out-side of the building (and which incidentally make it very difficult to reach the kitchen window over the sink). It is important to design the window reveals at an angle so that not too much light is cut off and to prevent too abrupt a contrast between the brightness of the window and the inside wall.

There are other ways of building with straw bales, either within a post and beam timber frame, or which use the bales as the structure which supports the building and which are rendered with lime ren-der inside and out. There can be difficulties with this approach, as the straw is susceptible to movement and will decay if moisture penetrates the render. This technique has been widely used in the south-west states of America, where there is a hot dry climate. Meanwhile, there is now a significant number of straw-bale houses in Britain, and one building firm that specializes in this form of construction.

The construction

The foundations and floor were made by a simple and cost-effective construction: a 200mm-deep reinforced concrete slab on 100mm expanded polystyrene insulation on 150mm of crushed stone, thickened to 500mm below ground level at the edges. The slab was powerfloated when laid to provide the finished floor, saving the more usual screed and tile or lino finish.

The conventional timber wall panels of vertical studs at 600mm centres support a roof of composite rafters made up on site from 'six by two' and 'four by two' flanges with short lengths of ply forming the web. This is again effective in supporting the heavy grass roof, and economical. The 400mm depth of the rafters is filled with cellulose fibre insulation made from recycled newspaper, held in place with breather membrane fixed to the top of the rafters. Above this are nailed 50x50mm battens, with a plywood deck fixed on top. This forms a 50mm ventilation gap to prevent the build-up of condensation below the roof membrane. This was delivered in one piece, lifted onto the roof and cut around the rooflight and other roof penetrations for the chimney and ventilation outlets, jointed using adhesive tape and weighted down with 100mm of turf taken from the site.

The rafters were fixed at the ridge and eaves on top of a strip of membrane to be lapped and taped to the vapour control and airtightness membrane incorporated under the rafters, and on the outside of the stud wall frames but inside the straw-bale insu-lation in the walls, to form a complete airtight and vapour-resistant layer. Unfortunately, the frame remained out in the weather for a year, during which time these strips became shredded by gale-force winds. However, the membrane was pieced together as best it could be and the building achieved a very good airtightness rating of 1.38 m^3/hr/m^2 of exposed construction at 50 pascals pres-sure difference between the inside and the outside.

Services are run within the stud frame, which forms a service void. The bales were stacked up and bonded using half bales made by pushing a steel nee-dle with baler twine through the bales. Small areas under windows were filled with straw by hand.

Horizontal 'ladders' of battens with 'rungs' of ply off-cuts were laid into the straw bales at two or three levels as the bales were laid. They tie the bales back to the frame and support vertical battens on a breathable membrane to improve the airtightness of the wall. The rainscreen cladding of Douglas fir boards, left natural without any applied finish, is fixed to the battens forming a ventilated cavity behind.

High-performance timber windows fitted with argon-filled double-glazing with super low E-coated glass and thermally broken edge spacers are fixed within ply boxes which support the windows and line the openings through the thickness of the straw insulation. The waterproof membrane is lapped under and over the window at the top and bottom. The window is sealed to the cladding with compressible, weatherproof, foam plastic tape.

The services

Heating is provided by a wood-burning stove with a back boiler, which provides almost all of the domestic hot water in winter. The single heat source provides comfortable conditions throughout the house because it is very well insulated. Hot water is provided in summer by a solar panel of evacuated tubes. It is stored in a highly insulated heat store with a rarely used back-up immersion heater for top-up heating in summer.

Extract ventilation is provided, with air being drawn in through controllable trickle vents in the bedroom windows. The bathroom is fitted with a continuous individual-room heat-recovery ventilation unit, with a high-efficiency DC fan rated at 2W with an automatic boost when the shower is on. The kitchen has a cooker hood which works by passive stack effect with a boost provided by a DC fan. Finally, air is continuously extracted at a low rate from the compost toilet.

The bathroom is provided with a conventional, but low-water-use WC, for those concerned about having a deep hole beneath them on the compost toi-

The compost toilet requires the occasional handful of wood shavings, and has a small extract ventilator so that it produces no odour.

let. Both it and the bidet discharge into the compost toilet chamber which is provided with a pump that removes liquid to a trench soakaway along with waste water from the sink, washbasin and washing machine.

Future developments

Nick would inevitably do a few, but not many, things differently if there were to be a next time: windows would be set further back from the face of the wall, with extended metal sills for greater durability; there would be fewer roof penetrations and an alternative product for the roof membrane.

Nick is now working on a development of this design which has all the structure and services on the inside of the airtightness and vapour control layer to avoid any penetrations of the layer. Meanwhile, he and Sheila are exploring different ideas for the conservatory at the back of the house and planning to link the wind generator and PV to the grid.

2.6 A modern, self-built house set in a West London conservation area

This house shows how a steel frame can create a lightweight, easy-to-build house which looks very elegant. A sunspace traps the sun's energy and provides additional living space.

Planning permission in a conservation area

This house comes as something of a surprise. Set between large Victorian villas and terraced cottages in a suburban neighbourhood in Ealing, it is uncompromisingly modern in appearance with a single pitched steel roof floating above a box covered with a gridded pattern all over and a double-height steel and glass winter garden to one side. The self-builders, John Brooke and Carol Coombes, never imagined at the outset that they would be living in a house which would be anything other than conventional in design with brick walls with small windows and a pointed roof on top; the site is in a conservation area, and they assumed that the planning authority would only approve a building which, in the jargon of planning legislation, "would preserve and enhance the visual amenity of buildings and character and appearance of the conservation area".

John and Carol interviewed "at least ten architects, most of them uninspiring" until they talked to two young architects who were the offspring of friends and just setting up in business after coming out of big, well-known, design-led practices. They proposed something much more adventurous. This design was supported by the conservation officer in the planning department, and eventually (after 15 months) the development control officers responsible were persuaded that a modern house would not compromise the quality of the environment of the conservation area. In fact the new house is a very positive addition to the scene, sandwiched as it is between a row of developer-built, 'traditional' houses of the worst kind and a 1950s bungalow.

A site at the end of the garden

The site was formed by combining the end of the garden of the large Victorian villa occupied by John and Carol and their three teenage children and a number of lodgers together with the site of a group of lock-up garages bought after a stand-off with the developer of a group of adjacent houses. John and Carol would not sell part of their garden but were able to buy the garages to make a viable single house plot. They do small-time property development and lettings, and intended to develop the site sometime with a house for sale, but when John took early retirement from the BBC they decided to build a new house for themselves instead.

The courtyard plan

The plan of the house is fairly simple in that it is set out on a grid and arranged according to a straightforward diagram: the accommodation is on one side of the site, and is as compact as possible, leaving room for a glazed winter garden on the other. The two-storey, enclosed side of the house contains an open-plan kitchen/dining/living-room and a study, utility area and toilet on the ground floor, and four bedrooms and two bathrooms upstairs. It is surrounded by thick walls to the north, east and west. The courtyard is flanked by a glazed wall to the house on one side, and a brick 'garden wall' to the other. It contains the staircase and a small mezzanine sitting area – both of which are galvanized steel. The effect is one of openness, sunshine and light. One of the consequences of the amount of glazing and the

FACING PAGE Above: First floor plan showing the stairs in the unheated courtyard. Below: Ground floor plan showing the large winter garden. OVERLEAF Left: The roof appears to float above a box with a gridded pattern of tiles. Right, top: The slender steel structure permits a very light and open interior. Bottom: The whole house opens onto the winter garden.

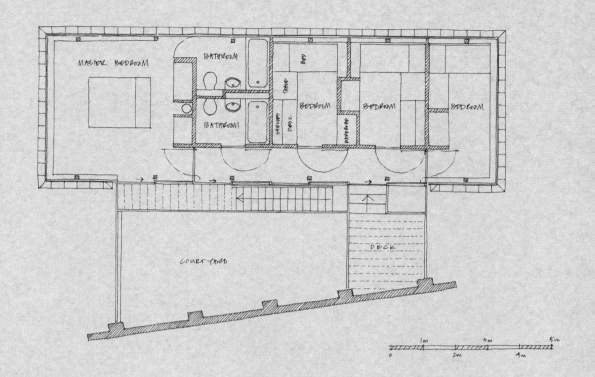

MASTER BEDROOM

BATHROOM

BATHROOM

BEDROOM

BEDROOM

BEDROOM

COURTYARD

DECK

0   1m   2m   3m   4m   5m

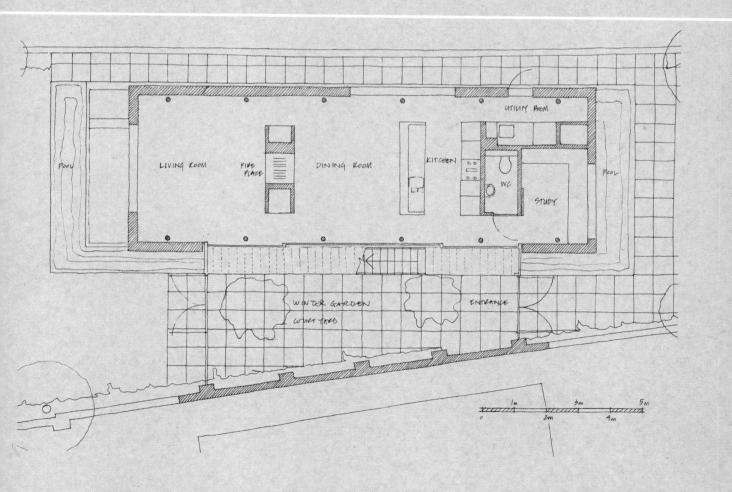

POOL

LIVING ROOM

FIRE PLACE

DINING ROOM

KITCHEN

UTILITY ROOM

WC

STUDY

POOL

WINTER GARDEN
COURTYARD

ENTRANCE

0   1m   2m   3m   4m   5m

position of the stairs is that life is lived in the public gaze, especially at night, so ingenious arrangements of blinds have been installed in the bedrooms and to the front of the courtyard.

Modernist self-build

The roof appears to float above a 'box' clad with ceramic tiles, with a continuous strip of glass between. The tiles form a very precise grid, which emphasizes the modular nature of the design and which is punctuated by a picture window to the living-room overlooking the back garden and a slot to the study overlooking the drive at the front. This box in turn appears to 'float' above the garden, and pools around the building emphasize this. The paving in the courtyard extends beyond the glass walls which enclose the courtyard; this reinforces the impression of the courtyard as an enclosed part of the garden. The effect is precise and modern, and in contrast to the somewhat folksy appearance of many self-build houses.

A self-builder's 'road map'

The avoidance of bricks and blocks and the use of lightweight construction on a modular grid with a steel frame with timber joists forming the floors and roof and composite I-beam studs in the external walls came about because John and Carol were going to build the house themselves. The modular grid gave John a 'road map' to find his way around the layout of the building. Frame construction is adaptable, because none of the walls carries the weight of the building; they can be relocated at will, and doors and windows can be created in any position. This has allowed the division between the house and the courtyard to be formed of fully glazed sliding doors on both floors. The steel frame, formed of standard sections, structural hollow sections (SHS) and parallel flange channels (PFC), is light in weight and slender. Steel is also a very adaptable

building material as it can be formed into any shape you wish. The steel frame is positioned inside the external walls to avoid heat being conducted to the outside through the steel forming a cold bridge losing heat and creating a condensation risk.

Trouble with foundations

The ground conditions of heavy clay with a number of trees nearby suggested piled foundations. This form of foundation can be advantageous for self-builders as it can significantly reduce the amount of heavy digging and concrete laying required, provided that reinforced concrete beams linking the piles just below ground level known as ground beams can be avoided. In this case some ground beams were required, and in addition the potential simplicity of piled foundations was compromised by constructing a concrete trench to gain access to extensive and essential storage space below the house. One of the worst moments of the

ABOVE: Top: The light steel frame is filled in with solid timber floor joists and I-beam studs in the walls.
Bottom: The house is raised above the ground, forming a useful storage space below the ground floor.
FACING PAGE Detailed section showing I-beams in wall on steel frame with cellulose
fibre insulation and rainscreen tiled cladding, forming a breathing construction.

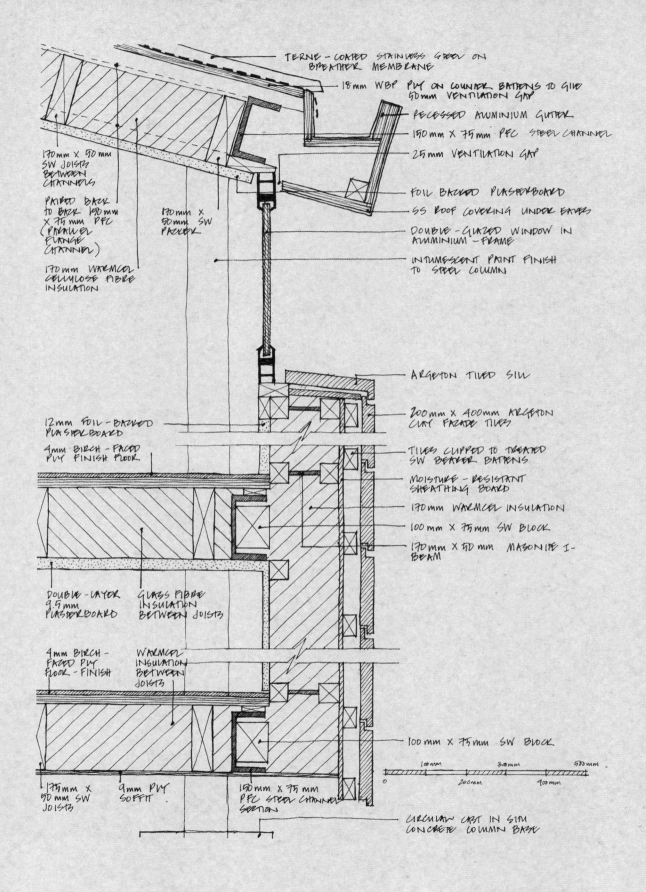

TERNE - COATED STAINLESS STEEL ON
BREATHER MEMBRANE

18mm WBP PLY ON COUNTER BATTENS TO GIVE
50mm VENTILATION GAP

RECESSED ALUMINIUM GUTTER

150mm X 75mm PFC STEEL CHANNEL

25mm VENTILATION GAP

FOIL BACKED PLASTERBOARD

SS ROOF COVERING UNDER EAVES

DOUBLE - GLAZED WINDOW IN
ALUMINIUM - FRAME

INTUMESCENT PAINT FINISH
TO STEEL COLUMN

ARGETON TILED SILL

200mm X 400mm ARGETON
CURT FACADE TILES

TILES CLIPPED TO TREATED
SW BEARER BATTENS

MOISTURE - RESISTANT
SHEATHING BOARD

170mm WARMCEL INSULATION

100mm X 75mm SW BLOCK

170mm X 50mm MASONITE I-
BEAM

100mm X 75mm SW BLOCK

CIRCULAR CAST IN SITU
CONCRETE COLUMN BASE

170mm X 90mm
SW JOISTS
BETWEEN
CHANNELS

PAIRED BACK
TO BACK 190mm
X 75mm PFC
(PARALLEL
FLANGE
CHANNEL)

170mm WARMCEL
CELLULOSE FIBRE
INSULATION

170mm X
50mm SW
PACKER

12mm FOIL - BACKED
PLASTERBOARD

4mm BIRCH - FACED
PLY FINISH FLOOR

DOUBLE - LAYER
9.5mm
PLASTERBOARD

GLASS FIBRE
INSULATION
BETWEEN JOISTS

4mm BIRCH -
FACED PLY
FLOOR - FINISH

WARMCEL
INSULATION
BETWEEN
JOISTS

175mm X
50mm SW
JOISTS

9mm PLY
SOFFIT

150mm X 75mm
PFC STEEL CHANNEL
SECTION

0   100mm   300mm   500mm
            200mm   400mm

build was when the formwork for this trench burst when the concrete was being placed, almost sending quantities of unmanageable concrete into the trench. This was narrowly avoided, whilst the formwork, which had been constructed out of oriented strand board (OSB) rather than the more expensive but stronger plywood, had to be repaired and strengthened, all while the ready-mix concrete driver was pushing to complete pouring the remainder of the load because his truck was blocking the road, much to the dismay of the local residents and police.

The self-build process

John worked full-time on the project, acting as labourer to his full-time carpenter Terry, while Carol was in the old garden shed (which acted as site hut), on the phone for a large part of the day organizing supplies or down the road placating drivers whilst lorries blocked the road making deliveries. John also directed operations, and made sure that the assembled experts knew what they had to do – and that they got paid every week. There are some advantages to being an inexperienced self-builder: you take nothing as read and seek the right advice, unlike experienced builders who are used to making all sorts of assumptions about how to do things, including techniques with which they may not be familiar, such as piling or steel roofing – and some of their assumptions may not be correct. There are limitations of course, such as not having the experience to know when to take shortcuts, and to know the level of accuracy appropriate to different operations. John describes setting out the building ready for piling as one of the scariest moments. Although there is a moderately high degree of tolerance at this stage, the important thing is that the frame is accurate, level and square, as otherwise everything that follows becomes a real problem. Also critical was setting the level of the drains so that there was sufficient fall to drain towards the connection with the drains serving the old house.

It can work both ways, of course: John recounts how he found that one of the steel fixers had found it awkward to get the drill in to make holes for the resin anchors that hold down the steel frame to the roof of the winter garden, and had just stopped drilling when it became difficult, leaving the anchors only half as deep into the concrete padstones as they should have been. There was therefore a risk that the roof would fly off in a high wind. With some lateral thinking, a way was found to drill the holes to the correct depth and so to ensure that the bolts were adequately glued with resin into the concrete supports. Generally, however, everybody from the truck drivers and delivery people, equipment hire company, builder's merchants and building control inspector, were extremely interested and helpful.

The day started with a get-together in the shed at 7.00 to make sure that the necessary information and materials were to hand. The packages of information from the architect were generally clear and available in good time; however the structural engineer used all sorts of abbreviations which were completely obscure to the uninitiated. As is so often the case, a great deal of effort was expended chasing suppliers; and as is also often the case, it was suppliers abroad who were the most reliable (the ceramic tiles from Germany and the perforated steel decking to the stairs from France arrived complete and undamaged on the appointed day). Co-ordinating the supply of the glazing was difficult, because there are many sizeable double-glazing units in the house, all of which are made of glass which is toughened for safety and laminated for security, and which all had to be measured on site before ordering. This meant that the frames had to be in position before manufacture could commence, which took three months. Meanwhile, the building was open to the elements and not yet secure.

Moving into an unfinished house

The workforce was increased from time to time by children on holiday from university and others, and some operations such as the glazing were

PREVIOUS PAGE The winter garden is an exotic extension to the house.
FACING PAGE Cross-section showing ventilation through the undercroft into the courtyard.

subcontracted. The building process started with demolishing the garages in August 1999. As is usual in the early stages of building, the structure proceeded relatively quickly and the roof was on by December. The family moved in on 1st October 2000 with the house unfinished, surrounded by mud and equipped with a Portaloo and a Camping Gaz stove but no front door; they had to wash their hair under a tap in the garden before going off to work.

The five of them could impose on friends no longer, having moved five times over the last year. The floor tiling went down in May 2001, which has been considered the completion date – although there is really no such thing as a time when a house is complete – and since then John and Carol have established a beautiful new garden which extends into the winter garden, and they have also built a garage – which in the way of many garages does not house a car, but is a studio for Carol who is a painter.

The cost of the original house was £310,900, which John and Carol consider good value although this is slightly more than twice the original budget.

The environmental strategy

The house is designed with small openings on the north to reduce heat loss, and with the winter garden on the south, which is double-glazed and designed to capture passive solar gains. The winter garden is provided with manually controlled vents at the top to prevent overheating. Cool air is drawn in from the undercroft below the house over the pool, which also keeps the humidity up for the benefit of the plants. However, it occasionally gets too hot on the sunniest of summer days, especially when the vents are shut because nobody has got around to opening them, or if everybody is out. The walls and roof have a high level of insulation provided by 200-250mm of cellulose

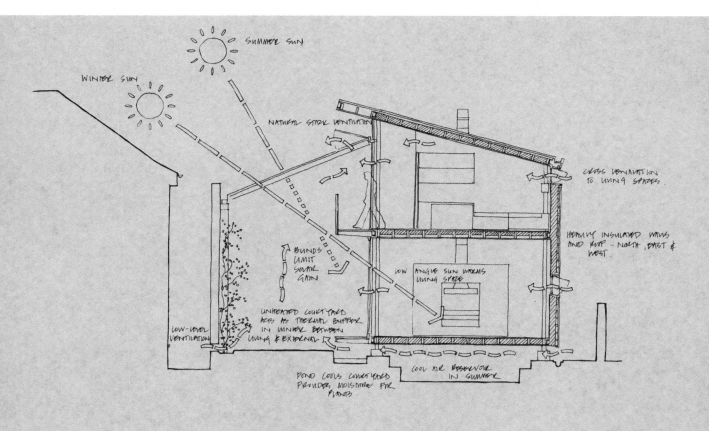

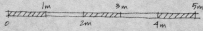

fibre. A gas condensing boiler drives an underfloor heating system with a sophisticated 5-zone control system (which suffers from over-sophistication – changing the batteries in the five thermostats is more trouble than it's worth). One sophistication that has worked well is the central vacuum system: just plug your hose into the outlet in each room – no more lugging the Hoover up and down stairs. The light fittings on the columns had to be purpose-designed and commissioned by John and Carol as no suitable designs at a reasonable price were available in this country. The bespoke design proved economical, as well as fitting in perfectly with its position, even if the wiring was awkward to thread through the steel structure to the fittings.

Time for some improvements

John and Carol are very at home in their new house, and now feel very comfortable with its modern design, even though it was not what they had in mind at the outset. They put their trust in their designers, and have had to find out about 'the rules of modernism', as John puts it, although inevitably there are a few things they would do differently – and now four years after moving in they are about to put some improvements in place. The amount of planting in the winter garden gives a wonderful feel of living in a garden, but has limited the usefulness of the space, especially when full of people at a party; the cooker hood is too far away from the cooker (so that it does not project below the line of the kitchen cabinets and spoil the clean lines of the kitchen), and so cooking smells can be a nuisance upstairs; the automatic controls for the vents in the roof of the winter garden, which were (regrettably) omitted to save money, could be added at a cost of £2,000; nuisance from the noise of people on the steel treads on the staircase could be prevented; and fitting the seals around the doors to the rooms which were manufactured with grooves to take them would improve the sound insulation. In other respects, this innovative self-build house has worked very well, and John and Carol have created a beautiful, energy-efficient home which shows that a low-environmental-impact house can look any way you want – ancient or modern.

The main bedroom has an all-round view.

Left: The original house was stripped back to the timber frame and restored. Right: The recent bedroom extension is built using hemp.

2.7 An architect uses hemp to build an extension

This sizeable extension demonstrates the use of hemp as a low-embodied-energy alternative to conventional masonry construction. It is an example of the application of green building techniques to an extension to an existing house.

Moisture and timber buildings

Ralph Carpenter is one of two partners in Modece Architects, an architectural practice in Bury St Edmunds that has carried out a wide range of thoughtfully designed, residential and social building around the area. He bought a house built in 1600 in a nearby village when he moved to the area in 1984. The front of the house was an apparently solid brick wall, which turned out to be a relatively recent addition applied to the front of the ancient timber-framed building behind, a practice common in this part of the world as elsewhere in Britain. This was a consequence of changing fashion, which suggested that a brick building was more solid and expressed the wealth of

the owners – an attitude which still has resonance with many people today. In fact, a brick wall is not weatherproof and lets water through the joints between bricks in driving rain, and is also susceptible to rising damp if not prevented by an effective damp-proof course. The brick wall around Ralph's house was particularly leaky because metal reinforcement had been incorporated in the horizontal joints which had corroded over the years and swelled as it had rusted, causing large cracks in the wall. Any moisture passing through the brick wall would be trapped against the timber frame, which would then tend to decay.

Timber will last indefinitely provided that it is not permanently damp; it will not rot provided that when it gets wet, on the exterior of a building for instance, it can always dry out again when the rain stops. Rot occurs for example when joists are built into damp brick walls or when timbers stand directly on the ground. In Ralph's house, the brick wall had to come down, as did the primitive lean-to on the back of the house which enclosed the kitchen and bathroom.

Reconstruction

So with the house stripped back to the original oak frame, reconstruction and extension could begin. The existing frame was insulated with mineral wool quilt with plasterboard on the inside, and cement render on galvanized steel mesh on the outside, and the windows were replaced with double-glazed timber windows. A new extension housed a kitchen and a well-glazed, east-facing, double-height dining area with a bathroom and storage space upstairs opening off a gallery. The extension was constructed using 150mm softwood studs in place of the old oak frame. This created a comfortable and at the time (20 years ago), an energy-efficient, three-bedroom family house.

Hemp for building

When the time came for a further extension, this was a clear case for using the construction method that Ralph had developed, using hemp on a range of fifty or so new and refurbished buildings since 1998. Hemp stalks are mixed with hydraulic lime to produce a stiff mixture which is packed into a plywood formwork between a timber frame to form a wall, which is finished with lime render on the outside and lime skim coat on the inside. Loose hemp fibre is packed into the roof, and a hemp/lime mix is used for the ground-floor slab. Ralph saw the potential of using natural fibres in building when he visited France with his brother, who was in a firm of agricultural merchants originally founded by their father. The firm was faced with the need to diversify in the face of restructuring of agriculture, and developed a business processing and supplying hemp for animal bedding and as fibre for the automotive and paper industries – and as insulation for buildings. 3,000 hectares of hemp is grown locally to the processing plant in East Anglia, generating new activity in the rural economy. A patented process of 'mineralization' to improve the durability of the hemp was developed in France. However, the plant in France has recently been closed down and so Ralph is using natural unprocessed hemp on an experimental basis for the latest project – a headquarters office and education

facility for a small environmental charity. The mineralized material has been used for 15 years or so in France and has been shown to be durable – it remains to be seen whether the untreated material is equally durable.

The advantages of hemp construction

The principal advantages of this form of construction can be summarized as follows:

- An extremely low amount of energy is required to produce the hemp material, and the emissions of CO_2 from the manufacture and curing of lime are very much less than Portland cement, which is now one of the principal sources of greenhouse gases in the earth's atmosphere

- The hemp/lime material is very porous to moisture vapour – it is a so-called 'breathing' construction. This has the effect of ensuring that condensation cannot build up in the construction, creating a risk of decay and reducing the thermal performance of the construction

- The hemp construction has a significant thermal capacity, which tends to reduce the risk of overheating in summer and slows down the rate at which a building cools down when unheated in winter

- A significant hemp industry would create employment and activity in the rural economy

- The construction is composed of renewable natural materials which do not pose potential hazards from toxic by-products. However, lime is an extremely caustic material which must be handled with care

- Hemp construction materials can be disposed of without harm when it is no longer useful – the timber can be reused, and the hemp/lime mix ploughed into the earth as a soil improver

Detailed section showing hemp used for the floor, walls and roof.
Note also that there is no damp-proof course membrane in the ground floor.

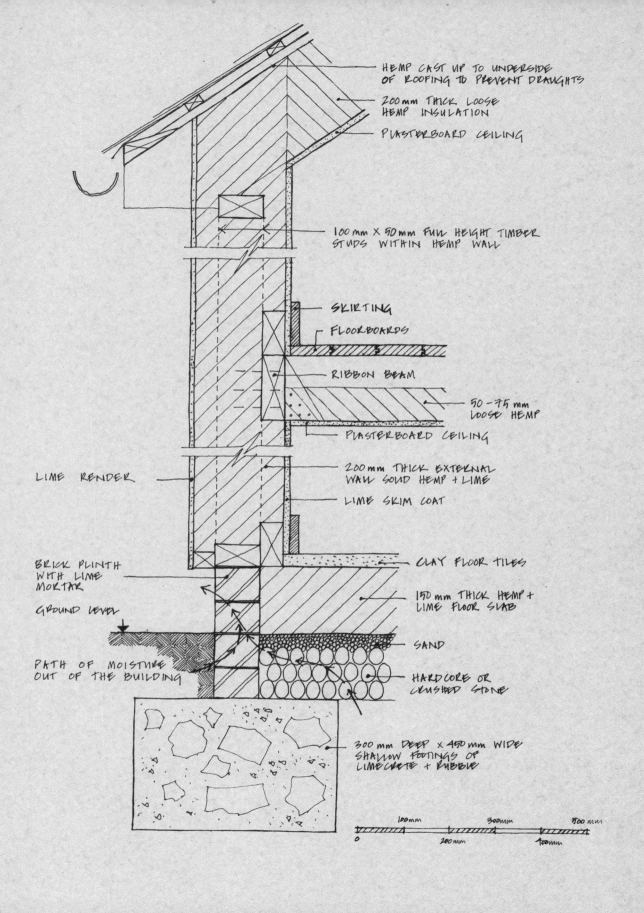

HEMP CAST UP TO UNDERSIDE
OF ROOFING TO PREVENT DRAUGHTS

200mm THICK LOOSE
HEMP INSULATION

PLASTERBOARD CEILING

100mm X 50mm FULL HEIGHT TIMBER
STUDS WITHIN HEMP WALL

SKIRTING

FLOORBOARDS

RIBBON BEAM

50 - 75mm
LOOSE HEMP

PLASTERBOARD CEILING

LIME RENDER

200mm THICK EXTERNAL
WALL SOLID HEMP + LIME

LIME SKIM COAT

BRICK PLINTH
WITH LIME
MORTAR

CLAY FLOOR TILES

GROUND LEVEL

150mm THICK HEMP +
LIME FLOOR SLAB

SAND

PATH OF MOISTURE
OUT OF THE BUILDING

HARDCORE OR
CRUSHED STONE

300mm DEEP X 450mm WIDE
SHALLOW FOOTINGS OF
LIMECRETE + RUBBLE

0   100mm   300mm   500mm
        200mm   400mm

Left: Walls are constructed by packing a mix of lime and hemp into formwork to fully enclose the timber frame.
Right: The hemp/lime mix is built up to the underside of the roof, which forms an airtight construction.

The results of monitoring hemp homes

Ralph first used the technique on a refurbishment project in 1995. 50mm of hemp was applied to the inside of the existing brick walls and to infill the old oak-framed external walls and partitions of the rear of the buildings. The method was tested on a project to build two affordable homes for rent as part of a larger development for Suffolk Housing Society. The homes were completed at the end of 2001, and the construction was monitored in detail by the BRE which compared the method with two adjoining conventionally built brick houses. They concluded that for the hemp houses:

- Strength and durability were equivalent to conventional construction. The hemp houses suffered some shrinkage cracking due to drying out. One of the consequences of using lime is that it requires a long time to dry (a 200mm thick wall as used in the trial takes about 8 weeks in summer) and will never dry out during winter. This means that construction should take place between March and September, which can be a severe limitation. The micro-cracking is sealed by an application of lime wash which is drawn into the cracks by capillary action where it sets to form an integral part of the material.

- Thermal performance was better than the masonry houses, although the calculated U-values suggested that it would not be as good. The masonry houses had cavity walls insulated with 100mm of mineral wool fibre blown into the cavity, 50mm closed cell polystyrene under the concrete floor slab and 200mm mineral wool quilt insulation in the roofs. The hemp houses maintained an internal air temperature 1^0 or 2^0C higher than the masonry houses, and this was confirmed by a thermographic survey conducted with a camera which takes an image which shows the surface temperature of the building – the warmer the building on the outside during a cold night, the worse the insulation. Ralph believes that this is a consequence of the conventional construction, using modern hard and impervious materials, hard-baked bricks, cement mortar and gypsum plaster, never drying out – it is constantly damp due to condensation being trapped in the construction. This is in contrast to the soft porous hemp construction, which allows moisture vapour to pass through towards the outside where it is evaporated away.

- Sound attenuation was somewhat less than for a cavity blockwork party wall, although it was sufficient to comply with the Building Regulations. The hemp material is, however, a very good sound absorber which creates a comfortable acoustic environment inside a house.

- The houses were entirely weatherproof.

- They cost more to build, as a consequence of the longer time required for the labour-intensive process of packing the hemp into the formwork to form the walls. There is also a learning

Left: A new timber frame extension houses the kitchen, dining-room and bathroom.
Right: The new principal bedroom looks out over the Suffolk countryside.

process which the contractor had to undertake, and the materials are relatively expensive as there is not a ready market established. The additional cost was estimated at 10%.

The details of hemp construction

The hemp construction has shallow footings 300mm deep made of 'limecrete' (lime with broken stone and brick as aggregate), as the hemp construction is relatively flexible and can tolerate a degree of movement without cracking. The ground floor is a hemp and lime slab 150mm thick on a base of 150mm crushed stone or hardcore. There is no damp-proof membrane, as the slab is water-resistant. Good ground drainage is important to keep the building dry. This creates a breathable, insulated floor which is then finished with a natural fibre carpet or clay tiles laid in sand to retain its breathable properties. The hemp walls are built on top of a brick plinth made with lime mortar, which raises the hemp above the splash zone at the foot of an external wall. The timber frame is made from locally sourced timber and is a form of balloon framing – the frames are the full height of the building as opposed to the storey-height platform framing more common in Britain these days. The first floor is hung from a ribbon beam, a horizontal timber fixed to the two-storey-high studs. The hemp infill provides the necessary bracing to the frame and completely surrounds the studs, providing a consistent backing for the lime skim coat on the inside and lime render on the outside.

Building the new extension

Ralph designed the second extension to his house using hemp construction for a two-storey wing with a new master bedroom with a shower room and balcony which enjoys extensive views over the surrounding countryside above a garage and workshop. He had previously obtained planning permission for a home office in the garden, which he was able to amend to an extension. The building incorporates solar hot-water panels, home-made flat-plate panels designed using the guide from the Centre for Alternative Technology, and rainwater harvesting. Rainwater from the roof is filtered and stored in two large tanks with a total capacity of three cubic metres for use in WCs and the washing machine. In summer the supply can be topped up with water pumped up from a well if necessary (this is the part of the country with the lowest level of rainfall).

Ralph employed a carpenter to construct the timber frame, which he did quickly and accurately. Ralph carried out the preparation work and ordering in the evenings and at weekends. He employed a plasterer turned general builder for the finishing stages, which he carried out to perfection. Ralph fitted the roof insulation and roof tiles, doors and windows and fitted out the garage and workshop downstairs, leaving the upstairs to the professional. The extension provides a beautiful main room with a fantastic outlook which is very comfortable, not too hot in summer or cold in winter, and which needs minimal heating.

2.8 A self-builder develops a specialism in earth building

This project demonstrates another low-embodied-energy, heavyweight, alternative method of construction, with a long pedigree in parts of Britain. Here earth is used to build a modern but very sculptural house.

A self-builder discovers cob

Born and brought up in east Devon, Kevin McCabe returned to the area after living a few years in London. He was working as a jobbing builder laying patios and building extensions, and got a job working on the renovation of a farmhouse listed as of architectural and historic interest. The project was grant-aided by English Heritage, who required the foreman on the job to be qualified in the use of lime in building – for mortar, render, plaster and paint. Kevin went on a one-day course organized by the Society for the Protection of Ancient Buildings (SPAB), which is the largest, oldest and most technically expert pressure group fighting to save old buildings from decay, demolition and damage: the SPAB represents a practical and positive side of conservation. Amongst other things, Kevin was responsible for rebuilding a cob wall as part of the farmhouse refurbishment. Cob is the form of earth building which used to be the principal form of construction in parts of Devon. There are estimated to be around 20,000 cob houses still in use in the county, although the building technique has fallen into disuse over the last 200 years or so. Earth mixed with straw is built up in layers around half-a-metre high to form walls around 600 to 900mm thick, which are left to dry for a couple of weeks and then trimmed to form slightly tapered walls, which are then lime-rendered on the inside and outside.

Kevin went on a course organized by the Devon Rural Skills Trust, given by a 70-year-old who had learned the techniques of building in cob from his parents. Kevin rebuilt the wall, and found the work easy and very satisfying. Kevin is a serial self-builder, and has made a career of building or renovating houses. His fourth was a virtually derelict stone barn with planning permission for an extension in stone. Although Kevin could have used stone, in the light of his recent experience of cob construction, he decided to reapply for planning permission for the work to be carried out in cob – and so started a new business as 'The Cob Specialist', persuading planners and potential customers that building in traditional cob is not only possible but also a good idea. This has led to a series of cob-built extensions and new houses over the last ten years or so.

Balancing moisture

Earth building is a very low-embodied-energy way of building, and avoids the use of energy associated with the production of fired clay bricks and Portland cement, which is now one of the principal sources of greenhouse gas production. It is a method of building with a very high thermal capacity; it takes a long time to heat up the building, but stores a significant amount of heat within the structure, which creates stable comfortable conditions in winter because the radiant temperature of the walls is relatively high, and in summer the building remains cool inside during hot weather. Kevin is of the view that a cob house is particularly comfortable to live in because it maintains a humidity level slightly higher than conventional construction. Like Ralph Carpenter describing living in his hemp house, Kevin emphasizes the role of the balance of moisture in the building to explain the thermal performance of the construction in the face of the theoretically poor thermal insulation of the walls – a U-value of 0.45 W/m^2 ^{0}C for the thick walls of cob with no additional insulation.

FACING PAGE Above: Earth is mixed with straw, built up in layers half a metre high, and trimmed to form slightly tapering walls. Openings for windows are formed around ply formwork. A plinth of lightweight, insulating blockwork protects the base of the wall from moisture.
Below: Earth and thatch make a very satisfactory organic combination.

Thermal performance

Kevin's own house, completed three years ago, is designed to achieve the maximum Standard Assessment Procedure (SAP) rating possible at the time: 120. (The energy performance of buildings has been changed with the amendment of the Building Regulations that came into force in April 2006. The SAP is based on the energy cost of heating a house, now on a scale of 1 to 100, but the thermal performance is now linked to the CO_2 emissions produced to accord with European Union directives.) Kevin's house achieves this high rating by incorporating high levels of insulation in the floor (100mm polystyrene below the screed which incorporates the underfloor heating) and 125mm mineral wool slabs below the thatched roof. The thatched roof provides the deep overhang necessary to protect the walls, and a plinth of local stone protects the base of the wall by lifting the earth above the splash zone and incorporating a damp-proof course. The stone is backed by a highly insulated construction consisting of two skins of aircrete blocks with 300mm of polystyrene insulation between. The house has an 11 kW oil-fired condensing boiler driving underfloor heating to the ground floor and providing domestic hot water. The three bathrooms on the first floor are provided with a heated towel rail. The top floor was built without any heating, in the expectation that the combination of warm air rising from below coupled with the high level of insulation in the roof would provide comfortable conditions. However, it has proved uncomfortably chilly on a cold night; the suggestion is that the thatch lets the cold north wind through and so a hole was made through the cob to duct warm air from above the log fire in the living-room but this proved insufficient and so a radiator has been added to the heating system. Cooking is by oil-fired Aga, and a large open fireplace in the living room has a log fire most evenings in winter.

A modern cob house

The house occupies a site with magnificent views. Kevin built the house for sale, but decided to move in with his family when it was finished. He designed the house himself, with help from an architect – who was 'not too headstrong', as he puts it – and the structure was calculated by a structural engineer. The 340m² house is planned on three split-level floors linked by two spiral staircases made of solid cob winding around the solid cob flue. Kevin was able to avoid fitting fire doors to the staircase and create an open plan feel to the interior because the Building Control officer agreed that it was not necessary, as the top inhabited level was less than 4.5 metres above the highest ground level. All the internal walls are cob, creating a massive thermal capacity. There is a cob-built larder in the kitchen ventilated to the outside, which keeps food cool, but not too cold – spreadable butter, for instance. Although built in a traditional manner, the house is very light and sunny and modern in feel. The interior is very colourful, with lime wash on a lime skim coat on earth plaster on the raw cob walls. There are no straight walls, and the interior is very sculptural. The exterior is raw cob at present, a deep red in colour with the texture of straw and stones showing through, but Kevin intends to apply a lime render in the traditional manner.

Local materials

Materials from the immediate locality are used as far as possible: earth from the field outside with a clay content of around 20%, but with straw and some coarse sand and 20mm-diameter stone added to control shrinkage. For cob building, the earth can have a clay content of between 10% and 30%, and about half the soil in the surrounding area and in the county of Devon is suitable. The upper floors are timber, on joists on green oak beams with pegged joints. The green oak lintels to the inside of the window openings have shrunk as the timber has dried

FACING PAGE Above: The huge top-floor den has proved chilly on a cold night, and so an extra radiator has been provided. Below: The log fire warms the massive central chimney.

out and created gaps which surprisingly have led to substantial air infiltration through the three-feet-thick walls. Oak is used for doors and door frames – local oak where possible – and imported plantation-grown material from France where long, straight and knot-free material is required. Local chestnut is used for the windows. This is very similar to oak; although a little more expensive, it generates far less waste as there is much less sapwood, which is not durable. Chestnut is more stable and not as hard, which means that it can be worked more easily. However, as he says, Kevin is not 'religious' about the specification: the ground floor has Italian porcelain tiles, which look terrific.

A cob building business

Kevin just can't stop cob building, and has completed a three-storey music room, studio and cider house annexe, a workshop-cum-barn and a cob pizza oven. A domed sauna building is under construction. He is building a similar house for sale nearby which incorporates a number of improvements. Firstly, there is an enclosure around the spiral staircase to the top floor to improve the sound separation. The living-room is separated from the stair for the same reason, and the underfloor heating will extend to the shower rooms to make it comfortable on the feet. Cob is not a particularly cheap way of building, and probably costs slightly more than a conventional brick and block cavity wall – whilst the material may cost little or nothing, it is a labour-intensive process to mix it, place it and trim it to form walls, and the building season is limited to between March and September. This new house will cost around £250,000 to build, for a 250m² building. It is, however, a very low-energy way of building and creates comfortable buildings with high thermal mass – and very appealing sculptural forms. The planning authorities in this part of Devon have been supportive of a sustainable approach to building which is a development of the local vernacular, and there is a market for his high-quality, low-energy houses.

Top left: The next house under construction.
Bottom left: Straw, coarse sand and stones are added to earth from the field to limit shrinkage. This wall will be finished off with a lime render.

Designing a good house

This section outlines the process of designing a house, and shows how this process differs when designing a 'green' building. Christopher Alexander's 'Pattern Language' is a very useful way of thinking about the elements required in a house, and the application of the Pattern Language to a particular project is described.

Sustainability should become an essential ingredient of design

A large part of the business of sustainable or ecological design is about designing buildings properly: designing for real needs, in a proper relationship with the climate and surroundings, and constructed of quality materials. This means creating buildings that are useful and long-lasting, that can be adapted to changing needs and expectations, that are loved and cared for – not to forget minimizing harmful impacts on the environment.

It is crucially important that environmental issues are considered right from the start: it is no good merely adding technical fixes as you go along – which will cost more, and will never be satisfactory. If an environmental strategy is in place from the start, your building should have a much enhanced performance at little or no extra cost. We must get to the point where green issues are as much part of designing buildings as considering the structure or safety in case of fire – in the same way that it is automatic that fire safety is considered throughout the building process.

You may very likely be employing an architect to help with the designing and obtaining the necessary permissions, and it is vitally important that he or she understands your aims and has the experience necessary to put them into practice. Ways of ensuring that the people that you employ will be able to realize your vision are discussed in the next chapter.

The process of design

The process of design is complex and difficult to define, but can be summarized as considering the many factors that a building has to take into account – practical needs, aesthetic judgements, personal preferences, costs, regulations, environmental impacts, structural adequacy, comfort – and manipulating them until they are in proper balance.

This process proceeds from the general to the particular, just as the patterns of the Pattern Language do.

Each aspect is considered in turn in the most general terms, and an initial synthesis is made which defines the overall limits of the project. With the broad outlines established, you will move on to give some definition to the outlines, to begin to quantify sizes and costs to the point of having an initial design. It is an iterative process, with more information and detail added at each turn of the circle. The next iteration will define matters such as the materials to be used, and will be firm and detailed enough to obtain a planning permission. Another turn of the circle will provide more detail on the structure, for example, and provide enough detail about the construction for Building Regulations approval to be sought. This will then form the basis for detailed orders for materials and components which define the size, specification and colour of the parts of the building. The design process continues throughout the construction period, with detailed decisions being made from day to day about how parts join together and what equipment to order, for instance. Conventionally the builder makes many of these decisions – sometimes consulting you first and sometimes doing what he thinks best, sometimes making the right choice and sometimes not.

As a general pattern emerges, opportunities are revealed for the next range of decisions, which when taken allow yet further developments to flow from them. It is an organic process and does not rely on the ability to conceive an entire, perfect concept at the outset.

Designing a green building

What differentiates the design process for a green building from any other is that the issues of siting, house layout, construction, structure, materials specification, heating, ventilation, landscaping and so on are all looked at from the perspective of reducing energy consumption, emissions, water consumption and waste, avoiding threats to health and allowing for adaptability in addition to the conventional aspects of cost, appearance, strength and so on. These issues have to be assessed at all stages – it is no good if, after your taking care to specify a certain detail or product, a contractor or subcontractor goes down the road and buys something else which is not environmentally sustainable because it's easier or cheaper.

This adds yet another layer of concern to the already complex process of design. It is also necessary to take care over design and specification, because standard tried and tested solutions are often not available, and low-environmental-impact buildings may have additional systems to design for water recycling, rainwater harvesting, wind generation or solar power.

The initial idea

The first idea may be no more than an estimate of the size of the building and a scribbled diagram showing how it might fit on the site, considering its shape, slope, outlook and access. Consider orientation, which will affect the degree of passive solar gains that you will be able to make use of. This initial analysis may have to be applied a number of times to different sites as the process of site acquisition proceeds. At the outset you will have to consider what the main aims of the project are – how important is cost, and how much you want to spend; what is your feeling about appearance (modern or traditional); whether you have a preference for a particular way of building (e.g. in timber, because it has a warm appearance or because it is relatively easy to work with hammer and saw). Visit completed buildings, and measure rooms which you are familiar with when planning the house.

Developing the diagram

Once there is some certainty about the site and the budget, diagrams can be refined to show the spaces required and how they relate to one another. What goes upstairs and what downstairs, how the inside spaces relate to outside spaces, where should the windows go, whether there should be balconies, a patio, a veranda, where are the views and existing

trees. You should make an initial assessment of the environmental strategy concerning siting, layout, materials, energy, heating, hot water, ventilation and renewable energy. You will have to consider the basic choices for the structural system and form of construction. Finally, you must make an estimate of the cost of this preliminary design.

Adding the detail

This can now be elaborated by revisiting these issues and firming them up, adding dimensions, making a preliminary layout for critical areas such as kitchen and bathroom so that you can allocate the right amount of space, and then accurately plotting the building on the site, checking overlooking and the relationship of the inside to the outside, considering new dimensions to the building – what will it look like in elevation, what materials, where are the openings. Consider the building in section to create volumes, not just spaces; stairs, galleries, double-height rooms and so on. Consider the structural alternatives in more detail and make choices for the construction of the floors, roofs and walls. Make initial choices for materials, particularly where they affect the appearance of the building, which will be a matter of interest to the planning authority; and for the heating and other systems in the building. Check the Building Regulations for the rules for Means of Escape in case of fire, energy performance, and other requirements. Pull all this lot together and check the cost estimate again.

Submitting a planning application

Once you are satisfied that the design achieves your objectives – provides the necessary accommodation in rooms that are delightful, light with good views and that the building has the required energy and environmental performance, relates to the site and its surroundings, meets the Building Regulations, can be built for the budget, looks great and possesses that quality without a name – then accurate drawings can be prepared and a planning application submitted. You may wish to refer to one of the other books around which give good advice on how to obtain planning permission.

The Building Regulations

Once a planning permission is obtained you will have to develop the design in more detail to show how it is constructed, to calculate the structure and energy performance, and to show how it complies with the Building Regulations. Detailed drawings will show how the main elements (the roof, walls and floors) are constructed, how they meet at the main junctions at the eaves and so on, and also how the special details where a roof abuts a wall or where a floor changes level are to be carried out. Everything now needs a dimension so that it can be built. You will need to obtain the technical details of materials, components and equipment, and assess their performance from an environmental point of view as well as the usual issues of strength, maintenance, cost and so on. Visit the showroom when deciding on bathroom fittings and so on. You will be obtaining quotations for alternative products and specifications and checking the cost again.

You will now be deciding who will be carrying out the building work – if you are planning to carry out most or some of the work yourself, employ subcontractors directly for each stage of the work, or get a general contractor to carry out all or most of the work. The issue here is that most contractors and subcontractors are not familiar with the issues involved in green construction. There is no point in specifying an airtightness membrane, for example, if the builder does not understand its purpose and how to install it and make the joints airtight. Employing a builder is discussed in the next chapter.

Designing has not stopped by this stage, because there will still probably be details of bathroom and kitchen layouts, boilers and ventilation systems, fin-ishes and colours that need to be decided. In some ways, if you are employing a builder it is desirable that as many decisions as possible can be made before starting building, because this makes the time and cost more certain. However, there are real advantages in leaving things until as late as possible so that you can see things *in situ* and make decisions with the building in front of you. This really only works if you are in full control of the process and if you are aware of the lead-in times from placing an order to materials arriving on site.

What is it that makes buildings survive well?

An important aspect of designing sustainable buildings which have a long-term future is to make adaptable to suit people's changing needs and aspirations, and ways of achieving this are discussed in Chapter 6 on building for longevity. Another essential ingredient is for buildings to remain wanted. They must be loved and cherished, for then they will be looked after, defended, changed and improved over the years. So green buildings need to remain useful – but what is it that will make them cherished and loved? Is there a special quality that makes a good building, and does it make some buildings survive while others are disposed of before their time?

The quality without a name

Christopher Alexander is an architect, teacher, thinker and writer who has developed a way of thinking about such a quality that makes places cared for, and makes them come to life. Although often 'we know what we like', we have not much exercised our minds as to why we like it. We may respond with pleasure to an old Cotswold village street with a great timber-framed barn, or a tile-hung cottage sheltering beside a copse, but we cannot put our finger on precisely what quality it is that moves us so. Alexander says: "It is easy to understand why people believe so firmly that there is no single, solid basis for the difference between good building and bad. It happens because the single central quality that makes the difference cannot be named." In his book *The Timeless Way of Building* he devotes himself to identifying that nameless quality, and at the end we realize that we knew all along what it was but were afraid to say so in case we seemed foolish in the eyes of the experts.

A Pattern Language

Having grasped the nature of the quality without a name, it is another thing to devise buildings that encompass it. Alexander and his team spent years researching and tabulating the universally recognized features that are common to the buildings we love. They uncovered a language of patterns from which these places were assembled. For instance, of one pattern, 'Light on Two Sides of Every Room' (number 159 in his book), which he claims "perhaps more than any other single pattern determines the success or failure of a room", he says: "Almost everyone has some experience of a room filled with light, sun streaming in, perhaps yellow curtains, white wood, patches of sunlight on the floor, which the cat searches for – soft cushions where the light is, a garden full of flowers to look out onto."

His next book, *A Pattern Language*, identifies and lovingly describes 253 such patterns (to be going on with, as it were, because the language grows as you use it and more patterns suggest themselves – it is an open way of thinking, not a closed system). Every time Alexander identifies a place that lives and takes us to the patterns that went into it, we recognize it. "If you can search your own experience," he says, "you can certainly remember a place like this – so beautiful it takes your breath away to think of it."

These patterns are the building-blocks that go to make the rooms, buildings, places, towns and cities around us. They make explicit those invariant features and attributes of places and buildings that work well and which we like, often without being able to articulate why. These patterns can be used to create an infinite range of examples, all different but all sharing those essential features that make them function well. A pattern language gives each person who uses it the power to create an infinite variety of new and unique places, just as everyday language gives a person the power to create an infinite variety of sentences.

Finding our own pattern language

In some form, every person has a dream to make a living world, a universe: whoever you are, you may have the dream of one day building a beautiful house for your family, a garden, a fountain, a fishpond, a big room with soft light, flowers outside and the smell of new grass. Alexander invites us to employ a pattern language whenever we aspire to build: "You can use it to work with your neighbours, to improve your town and neighbourhood. You can use it to design a house, for yourself, with your family . . ."

I commend these two books to you warmly as they are truly enlightening: *The Timeless Way of Building* for insights, and *A Pattern Language* as a practical reference book to help you make the right decisions in sequence, going from the general to the particular, from the broad sweep to the fine detail.

An example of the Pattern Language in action

Brian Richardson, former colleague and co-author with me of the original *Self-Build Book* published in 1991, recalls the design process that he and Maureen went through for their house in Herefordshire.

The Alexander books are a wonderful way of ordering your thoughts and deepening your insights, but are not a set of formulae that will of themselves produce design solutions. Having considered the appropriate patterns, you still do the actual designing yourself. Your decisions will come out of your own experience: Alexander helps you to recognize things you find you already know.

The books are delightful reading and easy to comprehend. The pattern language has the structure of a network – each pattern connects to the one around it. As you move through it you select appropriate patterns and develop them for your particular situation.

With the overall concept of the project delineated this way, construction can be embarked upon with confidence. As the building takes shape and work progresses, adjustments can be made by the self-builder to take into account unforeseen factors. This is in contrast with the orthodox architect-supervised building contract where the design is frozen at the moment the documents are signed and any subsequent variation provides excuse for a practically unstaunchable financial haemorrhage. The job thought out in the 'Timeless Way' can be a happy blend of careful planning and last-minute improvisation. True freedom!

To demonstrate its application, this is a brief selection from the range of patterns that Brian applied to his self-built environment in Herefordshire. He gives the name and number, and quotes the core description of the patterns used (not necessarily in Alexander's sequence), and then describes how he expressed it.

Old Age Cottage (155)

Old people, especially when they are alone, face a terrible dilemma. On the one hand, there are inescapable forces pushing them towards independence: their children move away; the neighbourhood changes; their friends and wives and husbands die. On the other hand, by the very nature of ageing, old people become dependent on simple conveniences, simple connections to the society about them. . . . Build small cottages specifically for old people. Build some of them on the land of larger houses, for a grandparent. . . .

In our particular circumstances, this pattern took a slightly different form. The 'small cottage' was already occupied by my widowed mother, and became a suitable habitation for an old person by our building the family home adjoining it. It was to our mutual advantage. She would be looked after in her old age and be able to stay in her own well-loved, cosy surroundings instead of going into a 'home'. We would have a superb building site with space, sun, a view and fertile land.

Settled Work (156)

Give each person, especially as he or she grows old, the chance to set up a workplace, within or very near the home. Make it a place that can grow

slowly, perhaps in the beginning sustaining a weekend hobby and gradually becoming a complete, productive, and comfortable workshop.

I was retiring from my profession; Maureen was getting under way with her new post-housewife one of papermaker. We could create a home and workshop together.

Building Complex (95)

Whenever possible translate your building programme into a building complex, whose parts manifest the actual social facts of the situation. At low densities, a building complex may take the form of a collection of small buildings connected by arcades, paths, bridges, shared gardens, and walls. . . .

The whole site, my mother's part and the portion ceded to us, would accommodate a complex of related buildings. The components would be her old cottage, garden shed and wooden garage in a much reduced and simplified garden; our small house, on a single storey and with wide doors for our own eventual old age; two workshops, one for home

building and maintenance and one for papermaking; greenhouses, woodshed and garden-tool store; and eventually, a pavilion summerhouse, all in a productive garden.

Site Repair (104)

Buildings must always be built on those parts of the land which are in the worst condition, not the best . . . On no account place buildings in the places which are most beautiful. In fact do the opposite. Consider the site and its buildings as a single living ecosystem. Leave those areas that are the most precious, beautiful, comfortable, and healthy as they are, and build new structures in those parts of the site which are least pleasant now.

The site in my mother's garden was the dominating factor in approaching the design. It was indeed the scruffiest part of the land and certainly would benefit from Alexander-type 'repair'. It had been used as a poultry run and was overgrown, but it had a splendid view from it. It was also blessed with an access road, electric power line and water main from

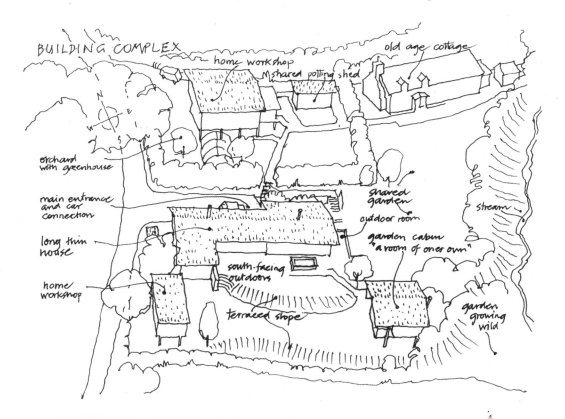

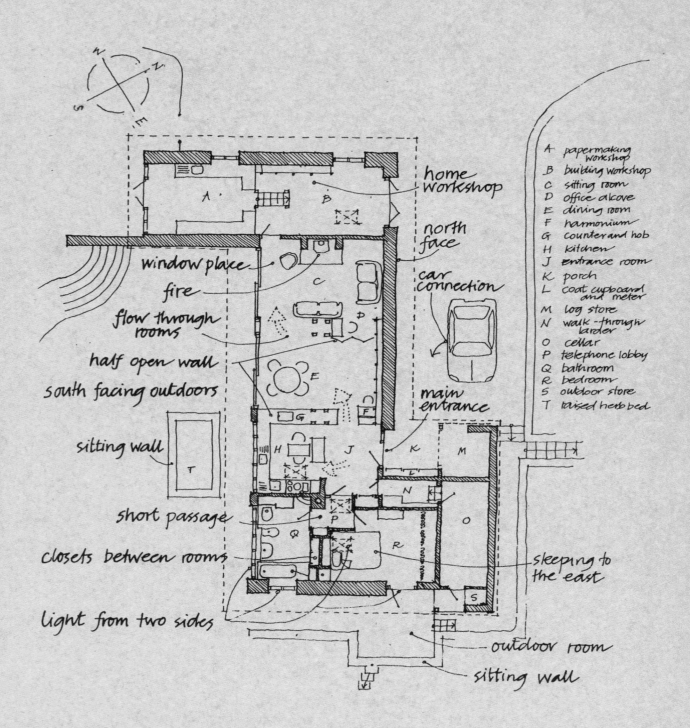

home
workshop

north
face

car
connection

main
entrance

window place

fire

flow through
rooms

half open wall

south facing outdoors

sitting wall

short passage

closets between rooms

light from two sides

sleeping to
the east

outdoor room

sitting wall

A papermaking
  workshop
B building workshop
C sitting room
D office alcove
E dining room
F harmonium
G counter and hob
H kitchen
J entrance room
K porch
L coat cupboard
  and meter
M log store
N walk-through
  larder
O cellar
P telephone lobby
Q bathroom
R bedroom
S outdoor store
T raised herb bed

Plan of Romilly noting several of the patterns used.

the north; distant view, summer breezes, sun from the south, a tree belt to windward, and a part falling away too steeply to use but with an intimate view to the east. A lot of time needs to be spent on site, soaking up its atmosphere and identifying its attributes. We had stayed often in my mother's cottage, but we camped out on the site a lot at this initial stage.

South-facing Outdoors (105)

Always place the buildings to the north of the outdoor spaces that go with them, and keep the outdoor spaces to the south. Never leave a deep band of shade between the building and the sunny part of the outdoors.

We paid attention to our neighbours' cottages, which seemed so well suited to their location. They were long and thin, built parallel with the contours at the north edges of their plots so that they formed a backdrop to their sunny gardens. We too decided to put the new house at the north edge of the undeveloped part of the garden my mother allocated to us, to leave the sunniest, most fertile, loveliest parts of the site to be gardened by her and us.

Sunny Place (161)

The area immediately outside the building, to the south – that angle between its walls and the earth where the sun falls – must be developed and made into a place which lets people bask in it . . . Inside a south-facing court, or garden, or yard, find the spot between the building and the outdoors which gets the best sun. Develop this spot as a special sunny place – make it the important outdoor room, a place to work in the sun, or a place for a swing and some special plants, a place to sunbathe. . .

The angle of the sitting-room window with the stone workshop wall makes just such a sun-trap on our garden terrace. It falls short of the full requirement of the pattern by being exposed to the southwest wind, though protected from the cooler and equally prevailing north-westerly.

Long Thin House (109)

In small buildings, don't cluster all the rooms together around each other; instead string out the rooms one after another, so that distance between each room is as great as it can be. You can do this horizontally – so that the plan becomes a thin, long rectangle; or you can do it vertically – so that the building becomes a tall narrow tower. . . .

We built long and thin, along the contour of the slope, running east-west, so that the rooms were shallow from front to back and were lit from windows on the south side, so that the main rooms would be sunny all day.

Flow-through Rooms (131)

As far as possible avoid the use of corridors and passages. Instead use public and common rooms as rooms for movement and for gathering. To do this, place the common rooms to form a chain, or loop, so that it becomes possible to walk from room to room.

The impression of space is enhanced in our house by not compartmenting the rooms into separate boxes but allowing them to interconnect in the sequence of the 'intimacy gradient'.

Main Entrance (110)

Placing the main entrance (or main entrances) is perhaps the single most important step you take during the evolution of a building plan. . . . Place the main entrance of the building at a point where it can be seen immediately from the main avenues of approach and give it a bold, visible shape which stands out in front of the building.

The entrance is visible and welcoming to visitors, with a sheltered place by it where one can pause outdoors before being admitted, where one can actually feel a first relationship with the house by grasping a post, or reaching up to touch the sheltering roof edge.

Entrance Transition (112)

Make a transition space between the street and the front door. Bring the path which connects street and entrance through this transition space, and mark it with a change of light, a change of sound, a change of direction, a change of surface, a change of level, perhaps by gateways which make a change of enclosure, and above all with a change of view.

As our east-west house is end-on to the lane down from the main road, it was easy to make a courtyard on the north side with, first, the entrance to the workshop off it and, at the more private end,

the house door. We found some nice stable paving bricks for a path along one side.

Car Connection (113)

Place the parking place for the car and the main entrance in such a relation to each other that the shortest route from the parked car into the house, both to the kitchen and to the living-rooms, is always through the main entrance. Make the parking place for the car into an actual room which makes a positive and graceful place where the car stands, not just a gap in the terrain.

Like it or not, the car is almost essential equipment in the kind of rural environment we were coming to. Besides being the only way my mother can be transported to the shops, the hospital and to visit friends, Maureen needs a car to conduct her business and we would have many car-borne visitors. A little parking and turning space at the end of the house, with the courtyard, can hold up to five cars when Maureen has students. Usually it is just our car that stands conveniently close to the front door without masking it.

Sheltering Roof (117)

Slope the roof or make a vault of it, make its entire surface visible, and bring the eaves of the roof down low, as low as 6'0" or 6'6" at places like the entrance, where people pause . . . roof edges you can touch. . . .

When approaching the house, it is good to have the feeling that it is a place of shelter. Even more than the walls, the enclosing roof emphasizes this character. The sweep of the roof, defining the shape of the house, and the overhanging eaves, throwing the rain clear, are very much symbolic of security and comfort. So our roof sweeps down over the front door, and Maureen has fashioned a little figure that peers down at the visitor as if it were saying, "Hello and welcome."

North Face (162)

Make the north face of the building a cascade which slopes down to the ground, so that the sun which normally casts a long shadow to the north strikes the ground immediately beside the building.

This car-space entrance-courtyard occupies the shady north side of the building, but needs sun let into it, so the roof slopes up gently from low eaves so as not to block out the sky.

Roof Garden (118)

Make parts of almost every roof system usable as roof gardens. . . .

We were very conscious of the beauty of the landscape we were building in, and wanted to intrude as little as possible. Our decision to build long and low below the skyline would help. We thought we must have natural stone walls, as our neighbours had, but we were left with a choice of roof to make.

A sod roof helps to relate the home to the topography around it. It would also provide us with a special bit of wild garden where flowers and grasses could grow untended – what the Friends of the Earth lovingly call 'unimproved pasture', because it does not get disturbed by cultivation and re-seeding. There would be lovely views down the Wye Valley from it when we wandered up there, but people in the valley wouldn't see much of us.

Intimacy Gradient (127)

Unless the spaces in a building are arranged in a sequence which corresponds to their degree of privateness, the visits made by strangers, friends, guests, clients, family, will always be a little awkward. . . . Lay out the spaces of a building so that they create a sequence which begins at the entrance and the most public parts of the building, then leads into the slightly more private areas, and finally to the most private domains.

Some more precise form now began to develop in our minds for the interior spaces. Inside the front door is a semi-public space with tiled floor, coat pegs, a ledge to put things down, a glimpse into the house of things to come – either kitchenwards, if the call is informal, or to the dining space for more formality. Beyond that is the more intimate sitting space around the fire, or in the other direction, through a buffer lobby, the lavatory and bathroom and the most private place, the bedroom.

Half-Open Wall (193)

Rooms which are too closed prevent the natural flow of social occasions, and the natural process of transition from one social moment to another. And rooms which are too open will not support the differentiation of events which social life requires. . . . Adjust the walls, openings and windows in each indoor space until you reach the right balance between open, flowing space and closed cell-like space. Do not take it for granted that each space is a room; nor, on the other hand, that all spaces must flow into each other. The right balance will always lie between these extremes: no one room entirely enclosed, and no space totally connected to another. Use combinations of columns, half-open walls, porches, indoor windows, sliding doors, low sills, French doors, sitting walls, and so on, to hit the right balance.

Our spaces are differentiated by structural frame posts, screens, counters, suspended shelves, but not usually by completely solid walls. The lighting at night is switched to allow the spaces to be articulated as pools of light.

Short Passages (132)

Long sterile corridors set the scene for everything bad about modern architecture.

We managed to do without any corridors, but where a ventilated and sound-blocking lobby is required between bathroom, bedroom and the living

rooms, we made it short, well lit and an interesting space in its own right. We put a telephone, some pictures and bookshelves in it.

Sleeping to the East (138)

Give those parts of the house where people sleep an eastern orientation, so that they wake up with the sun and light. This means, typically, that the sleeping area needs to be on the eastern side of the house; but it can also be on the western side provided there is a courtyard or a terrace to the east of it.

We wanted a special bedroom for ourselves, extremely private, compact in space but with lots of bookshelving and somewhere to hang our clothes. We would have glorious views from every other room in the house and thought there should be some place with a more cave-like, comforting atmosphere – and we would be in the room mostly at night anyway. We gave the bathroom the flood of sunshine and the view to encourage us out of bed in the morning, but the bedroom is just lit from the east, so the morning sun falls across the end of the bed. We are sheltered too from the buffeting of the prevailing west wind. A skylight over the bed gives a glimpse of the stars and illumination for daytime reading.

Bathing Room (144)

"The motions we call bathing are mere ablutions which formerly preceded the bath. The place where they are performed, though adequate for the routine, does not deserve to be called a bathroom."
– *Bernard Rudofsky.*

Concentrate the bathing room, toilets, showers, and basins of the house in a single tiled area. Locate this bathing room beside the couple's realm – with private access in a position halfway between the private secluded parts of the house and the common areas; if possible, give it access to the outdoors; perhaps a tiny balcony or walled garden. Put in a large bath – large enough for at least two people to get completely immersed in water; an efficient shower and basins for the actual business of cleaning; and two or three racks for huge towels – one by the door, one by the shower, one by the sink.

Alexander gives much significance to bathing – socially as well as personally important in his view – and we certainly enjoy bathing together. Our bathroom therefore needed to be spacious and beautifully lit and a congenial place to be naked in; quite the opposite situation of the all-too-common internal washing and defecating cupboard, artificially lit and ventilated, cramped and lined with cold surfaces. We instead gave the best, south-east corner of the house to it so we would have daylong sunshine from windows in both walls, cross-ventilation, a glimpse of the Brecon Beacons, enough room to move about from WC to bidet, from shower to deep

bath. We lined it with wood, hung it with baskets of plants, stacked it with magazines – and it was to be at least as nice as any part of the house.

Farmhouse Kitchen (139)

The isolated kitchen, separate from the family and considered as an efficient but unpleasant factory for food is a hangover from the days of servants; and from the more recent days when women willingly took over the servants' role. . . . Make the kitchen bigger than usual, big enough to include the 'family room' space, and place it near the centre of the commons, not so far back in the house as an ordinary kitchen. Make it large enough to hold a good big table and chairs, some soft and some hard, with counters and stove and sink around the edge of the room; and make it a bright and comfortable room.

The kitchen is in the middle of our house and much activity centres on it. It is near the entrance door, is surrounded by waist-high shelves and worktops, the walls are lined with open storage shelves and there is a central table. Cool stores on the north side open off it for food and drink.

Closets (Cupboards) between Rooms (198)

Mark all the rooms where you want closets. Then place the closets themselves on those interior walls which lie between two rooms and between rooms and passages where you need acoustic insulation. Place them so as to create transition spaces for the doors into the rooms. On no account put closets on exterior. It wastes the opportunity for good acoustic insulation and cuts off precious light.

All our rooms are separated from each other by thick walls containing built-in storage space; as well as being essential in its own right (Alexander suggests 15% or 20% of the house area for bulk storage), this serves to buffer sound.

The Fire (181)

Build the fire in a common space – perhaps in the kitchen – where it provides a natural focus for talk and dreams and thought. Adjust the location until it knits together the social spaces and rooms around it, giving them each a glimpse of the fire; and make a window or some other focus to sustain the place during the times when the fire is out.

The natural sequence in the long thin house is from kitchen to dining-room to sitting-room around the fire at the far end. There we have sitting space that can be grouped round the hearth or the window to the garden.

Light on Two Sides of Every Room (159)

When they have a choice, people will always gravitate to those rooms which have light on two sides, and leave the rooms which are lit only from one side unused and empty. This pattern, perhaps more than any other single pattern, determines the success or failure of a room. The arrangement of daylight in a room, and the presence of windows on two sides, is fundamental.

In the rooms at the corners of our house, the workshops and bathroom for instance, it was easy to put windows in adjoining walls. An advantage of a single storey and a low roof-pitch is that roof lights can allow light to fall in another direction than the windows, and we did this. Our experience of living in the house certainly corroborates Alexander's observation that "this pattern alone is able to distinguish good rooms from unpleasant ones". It is so nice in the parts of the house where we have exploited 'light on two sides' that I wish we had put in two or three more roof lights.

Window Place (180)

Everybody loves window seats, bay windows, and big windows with low sills and comfortable chairs drawn up to them.

The long south wall gives us the opportunity for low sills and sunny window places along it.

Connection to the Earth (168)

Connect the building to the earth around it by building a series of paths and terraces and steps around the edge. Place them deliberately to make the boundary ambiguous – so that it is impossible to say exactly where the building stops and earth begins.

The paving stones of the terrace are laid with spaces between for mosses and plants; eventually they give way to lawn. The indoor space flows imperceptibly into the outdoors, connecting house, garden and distant view into a whole.

Outdoor Room (163)

Build a place outdoors which has so much enclosure round it, that it takes on the feeling of a room, even though it is open to the sky. To do this define it at the corners with columns, perhaps roof it partially with a trellis or a sliding canvas roof, and create 'walls' around it, with fences, sitting walls, screens, hedges, or the exterior walls of the building itself.

As well as our main terrace, which is sunny but very open, there is at the east end of the house a morn-

ing suntrap. There we have cut into the bank a semi-sunken outdoor room, paved with brick – a lovely place to breakfast and enjoy a slanting glimpse of the distant view past the wood, completely sheltered from the wind. Opening out from our bedroom, this complements its cave-like quality and gives us the opportunity to sleep outdoors in the summer. It makes the house larger without making the garden smaller!

Sitting Wall (243)

Surround any natural outdoor area, and make boundaries between outdoor areas with low walls, about 16 inches high, and wide enough to sit on, at least 12 inches wide.

Our outdoor room is bounded by a sitting wall, popular at morning coffee time.

Home Workshop (157)

Make a place in the home, where substantial work can be done; not just a hobby, but a job. Change the zoning laws to encourage modest, quiet work operations to locate in neighbourhoods. Give the workshop perhaps a few hundred square feet; and locate it so it can be seen from the street and the owner can hang out a shingle.

The remaining part of our building complex is the home workshops. Where other people would need to devote space for bedrooms for their children or for visitors, we dedicated the equivalent space to workshops. Being at the extreme west end of the long thin house, they can be shut off at a narrow pass-door so that noise and smell do not enter the living quarters, and so that some separation of work and leisure can be attempted, if not totally achieved.

One end of the workshop is for Maureen's paper-making, the other (upgraded from the original appellation of garage – it soon became obvious that it was too precious a space ever to put a car or even my motorbike in) is for general building and maintenance.

The requirements of a workshop are somewhat similar to those of a kitchen: a lot of wall space for open shelving; waist-high benches for mounting tools and placing workpieces; good light from two

sides; everything to hand for the most common tasks. These needs we met, but we made it too small! So we built an extension, a freestanding timber-frame building that is a covered veranda. Our plan is for Maureen to expand further by taking over the general workshop too, which will then be replaced during the rebuilding of the tumbledown garage belonging to the old cottage.

Terraced Slope (169)

On all land which slopes – in fields, in parks, in public gardens, even in the private gardens around a house, make a system of terraces and banks which follow the contour lines. . . .

The building, cut into the bank, provides material for levelling a series of garden terraces down the slope to the south, so that when our eyes drop from the distant view, we can see our vegetables thriving. Screened by the bank is our row of compost heaps.

Garden Growing Wild (172)

A garden which grows true to its own laws is not a wilderness, yet not entirely artificial either. . . . Grow grasses, mosses, bushes, flowers and trees in a way which comes close to the way that they occur in nature: intermingled, without barriers between them, without bare earth, without formal flower beds, and with all the boundaries and edges made in rough stone and brick and wood which become part of the natural growth.

As well as the terraced slope, the garden takes its form from its location, sheltered by the woodland to the east, where, on the banks of a steep dingle, trees and plants grow naturally. High hedges break the wind from the north, and the wing of workshops encloses it on the west boundary.

A Room of One's Own (141)

No one can be close to others, without also having frequent opportunities to be alone. . . . Give each member of the family a room of his or her own, especially adults. A minimum room of one's own is an alcove with desk, shelves, and curtain. The maximum is a cottage. . . . Place these rooms at the far ends of the intimacy gradient – far from common rooms.

This pattern has to do with the privacy individuals need within a relationship. Maureen has her own domain as a working papermaker, and has even annexed an alcove in the living-room for specialist books and a second desk for business correspondence. But my drawing-board, files and writing materials used to get shifted about the house and the veranda, depending on the weather and the ebb and

flow of visitors. So we have built another outbuilding to complete the complex, with the double role of writing room and guest wing – a summerhouse pavilion in the wooded part of the garden with a little wood-burning stove and a paraffin lamp.

In 1997–1998 I set about getting that 'room of my own' using a windfall gift of second-hand timber from my son. I turned again to the Pattern Language, this time designing 'backwards' from the very sturdy and rather short baulks of Baltic Pine.

I drew heavily on the range of patterns occurring in the main house – emphasizing some, adding a few more. Particularly important was to site the new building correctly, remembering 'Site Repair . . . buildings must always be built on those parts of the land in the worst condition' (104) and considering 'Positive Outside Space (106) . . . make all the outdoor spaces which surround and lie between your buildings positive. Give each one some degree of enclosure. Surround each space with wings of buildings, trees, hedges, fences, arcades and trellised walks until it becomes an entity with a positive quality and does not spill out indefinitely around corners.'

The new building forms the east boundary of a courtyard open to the south and the valley below. It satisfactorily completed the composition of Building Complex (95).

The easiest way for an elderly person to build is in post and beam timber frame, as Walter Segal ordained. Foundations are particularly simple as the site does not have to be made level. So the pattern Terraced Slope (169) could here take the form of a retaining wall at the top of the slope acting as a landing stage to which the timber-framed building appears to be moored.

This automatically creates a void between the underside of the floor and the slope of the ground, which provides ample space for Bulk Storage (145), so difficult to find enough of . . . 'include a floor area at least 15 to 20 per cent of the whole building area.' It happens that the floor area of the cabin measured

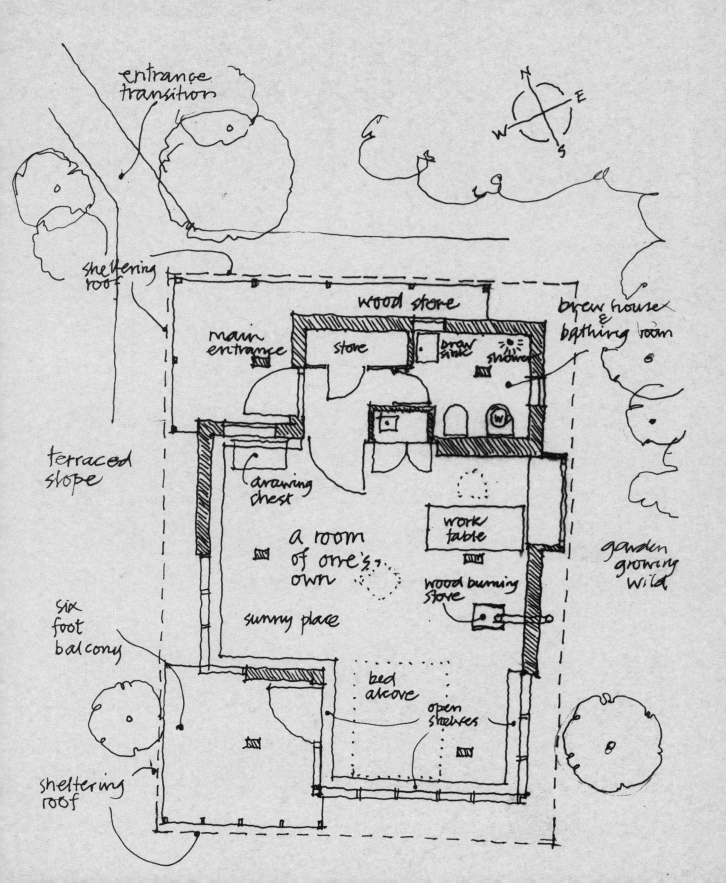

A ROOM OF ONE'S OWN

entrance transition

sheltering roof

terraced slope

six foot balcony

sheltering roof

wood store

main entrance

store

brew sink

shower

brew house & bathing room

drawing chest

a room of one's own

sunny place

work table

wood burning stove

garden growing wild

bed alcove

open shelves

Sheltering Roof (117) and Six Foot Balcony (167) respectively. The wide eaves control the flow of summer and winter sun into the building, which has windows all round, as in Light on Two Sides of Every Room (159) and makes the south-west corner an ideal Sunny Place (161). The spreading roof protects the front entrance porch 'Where people pause . . . roof edges you can touch (110)'; and covers the cantilevered balcony. This is two metres square – the minimum size that is useful for a sitting-out space: 'Whenever you build a balcony or porch . . . always make it six feet deep' (167).

The design that unfolded so naturally using the Pattern Language has produced a place that lives. I find myself continually gravitating to it, always with a tingle of enjoyment. I consider it the most rewarding project I have ever undertaken.

inside the external walls is 30 square metres, while the whole roof area measures a surprising 60 square metres. This is a result of exploiting the patterns for

How can you encourage others to think green?

You will have to involve a lot of other people in your project, many of whom will be sceptical about the idea of green building. This section outlines some ways of convincing them that green building is not something to be afraid of. It also describes how to go about employing designers and contractors who will help you realize your aims, and how to deal with subcontractors.

Influencing and employing other people involved in your building project

Designing and building a house is a complex process, and you will be relying on a range of other people for parts of the process. You may be obtaining finance from a bank or building society; planning permission will have to be obtained (and you may find yourself negotiating with the local authority planning officer); the interests of the neighbours may have to be addressed; and approval of the design, materials and structure will have to be given by the building control officer. You will have to influence all of these people and others to support your aims. You may be employing consultants, sub-contractors or suppliers to design or build all or part of the building, and you will have to ensure that they also understand your aims and have the knowledge and experience to turn them into reality.

The perceived risks of green building

Many of these people will have preconceptions about how to design and how to build. They may be wary of new techniques, materials and approaches. The banker may perceive a green building to be a greater financial risk, with higher costs and lower values and therefore be unwilling to lend; the planner and the neighbours may not agree that the benefits of building using materials with a low environmental impact such as timber outweigh the perception that a timber building would be inappropriate in a neighbourhood of predominantly brick buildings; the building control officer may be unfamiliar with the use of certain materials such as sheep's wool for insulation and may be wary of granting permission for their use. Suppliers will be used to using short cuts. For example, they may design the heating system using general rules of thumb, such as fixing larger pipes than necessary which may lead to a less efficient system; the

subcontractor, used to building as he has always done, may not appreciate the importance of making sure that the building is well sealed against air infiltration to reduce heat losses, and so not take the care necessary to ensure that gaps are properly sealed.

Green building also has advantages

You may find that you have to convince some of these people of your intentions, and to carry them along with your plans. There are of course a number of significant benefits of green building, which will improve the world we live in for everyone in an indirect way:

- Reductions in the use of fossil fuels to provide heating and electricity for homes will reduce emissions of CO_2, which is a major factor in global warming.

- This will also reduce other pollution, conserve stocks of fuel and make Britain less dependent on supplies of oil and gas from unstable parts of the world.

- A reduction in the use of water will conserve resources which are already under stress in parts of England.

- Better quality buildings with a long useful life will help to conserve resources and improve the economy.

Policies support green building

These measures are all underpinned by international agreements including the Kyoto Protocol and the Intergovernmental Panel on Climate Change, the IPCC, on CO_2 emissions. The European Union has been leading policy development in many areas, and the Directive on the Energy Performance of Buildings became effective throughout Europe in January 2006. These international agreements are reflected in British government policies, including national targets for sustainable development and the Sustainable Communities Plan. There are in turn regional targets and plans published by the Regional Development Agencies (the RDAs). Local plans for each local borough and district incorporate the regional targets, and are in turn translated into policies for support for energy conservation for example in each local government area.

> *A green building need not look much different from any other building. Buildings with low environmental impact can be designed to enhance any locality.*

As you can see, there is a plethora of policies at all levels. It is well worthwhile obtaining copies of the documents pertaining to your area and familiarizing yourself with the relevant points in order that you can quote them in support of a planning application, for example.

How to influence people

You may have to convince the financier by explaining the business case for green building and showing that it does not present unreasonable risks. There is a growing body of experience which can demonstrate the benefits and which suggests that green buildings can command up to a 10% premium in value. They will also tend to be an appreciating rather than a wasting asset. There is also a growing body of experience, particularly on the continent, of tried and tested techniques and products. If you can produce evidence of this kind, it will show that the risks are relatively low.

A green building need not look much different from any other building. Buildings with low environmental impact can be designed to enhance any locality, and there are built examples that can be used to persuade planning officers and adjoining owners that they have nothing to fear. Planners should be encouraged to support high-quality schemes as a means of improving standards by example.

Detailed technical information, test results, certification and calculation will have to be employed to demonstrate to the building control officer the safety and fitness for purpose of your proposed building and its methods of construction and materials. Again, building control officers should be encouraged to support proposals that go beyond mere bottom line compliance with the Regulations to demonstrate best practice in energy efficiency and high-quality construction.

It is vitally important to get people on your side. A good principle to adopt to help in this process is to develop a collaborative relationship as far a possible with the people you are dealing with; officials can provide a great deal of useful advice, and neighbours will be generally very supportive if they understand what you are trying to achieve.

Employing a designer

An important question is who is going to do the designing? You could do it yourself, which would be in the spirit of self-help, but there are many technical aspects that will have to be dealt with. You

Most people would be well advised to appoint an architect – but do not be downcast! You can still be the designer of your own home if you are clear about what you want and remain in control of the project.

imum energy consumption and low environmental impact.

You could decide to purchase a standard kit house or a standard design for a house. This may simplify the approvals process, but you then forsake one of the principal opportunities of building your own home – to have a house designed to your particular requirements and desires.

Whilst you may be able to rely on the house-building industry to produce a reasonable standard house (probably with not much delight), they will not be able to deliver a house with a good environmental performance.

Most people would be well advised to appoint an architect – but do not be downcast! You can still be the designer of your own home if you are clear about what you want and remain in control of the project. You will be getting professional help in getting it right and making sure that it is a quality building in harmony with the environment – so long as you get the right person.

Your designer will have to understand the principles underlying a sustainable approach to building and be committed to spending the time and effort necessary to getting it right and putting aside their prejudices. They can benefit from opportunities to work on cutting-edge developments, learn new skills, and acquire expertise that will put them ahead of the competition and open up new markets in the future.

Many professionals are still trained to believe that they know all the answers; they will tend to mystify their knowledge. You are looking for a designer who will explain the situation, help you to decide what you want and assist you in getting it. They should have experience of house building and of environmental design, and it is helpful if they are not too far away. There is a need for someone with experience and expertise in green building – and not that many people have it yet. The Association of Environment Conscious Building (AECB), has a useful register online of builders and suppliers, but also

may have the necessary experience of planning and building regulation and enough knowledge about construction and environmental design to be your own green architect, and you will certainly be in control of the process. Common sense will take you a long way, but drawings and specifications will be required to be submitted for approval, and whilst most officials will do what they can to be helpful, they do prefer in the end to deal with a technically qualified person who knows what is required and can produce the necessary information.

You could opt to employ an unqualified draughtsperson to turn your ideas into a design. They will tend to be limited in their capabilities and almost certainly inexperienced in designing for min-

designers with a particular interest in and experience of green building. Be guided by recommendation, visit some completed projects and talk to the owners, get references and follow them up. Decide what tasks you want them to be responsible for, interview more than one and go through these tasks and gauge their reaction. Obtain quotations for the work. Finally, make sure that you have a clear agreement on the scope of the work and the fee – it sounds obvious, but it is surprising how often this is ignored.

Employing builders

Many builders similarly believe that they know all the answers and that there is only one right way – their way – of doing anything. You are looking for a builder who will consider the situation, suggest ways to get what you want and assist you in getting it. There is again a need for someone with experience and expertise in green building – and not that many people have it yet. The AECB register will come in handy here. Again, be guided by recommendation, visit some completed projects and talk to the owners, get references and follow them up. Decide the scope of the work you want them to take on, interview more than one, and go through the job and gauge their reaction. Obtain tenders for the work. Finally, make sure that the scope of the work is clear and agree the cost – again, it is surprising how often this is ignored.

Controlling subcontractors

One particular problem that arises when you employ a general contractor is that you tend to have very little control over whom they employ as subcontractors – they may have no particular experience of environmental construction and/or have been to the 'we know best' school of building. There are ways of specifying which subcontractors should do the work, and they are often used if the work is of a specialized nature, but these arrangements are difficult for general building as contractors tend to have a group of subcontractors with whom they work on a regular basis; and they do not like having to work with people they do not know and trust, and the cost of the work will go up to cover the cost of this risk. If you act as a general contractor yourself, you will be able to employ all the subcontractors directly and vet them for their experience and attitude. There is a lot of work involved in employing subcontractors, of course, which you are paying a contractor to do, so if you do it yourself you will be saving maybe up to 20% of the cost of the job. You also need some experience to negotiate a fair deal and to make sure that the subcontractors' work is up to standard.

One solution that is used is to get your architect to employ the subcontractors on your behalf and to rely on his or her resources of time and experience. This is to add to the architect's role during construction, which is normally that of 'contract administrator': essentially the role of project manager administering the contract, monitoring the progress of the work, valuing the work for purposes of payment and signing the Completion Certificate which confirms that the work has been completed according to the drawings and specifications. This certificate may be required for mortgage purposes.

Environmental issues and the site

This section considers the environmental issues that you should consider when choosing or developing a site. You will need to consult other references for advice on planning and other matters which apply to any development.

Access to amenities

One of the principal sources of greenhouse gas emissions in Britain and elsewhere is the transport sector, and it is this sector where emissions are rising faster than elsewhere. An important part of reducing one's carbon footprint (the measure of the amount of CO_2 emitted to the atmosphere as a result of household energy use, transportation and waste disposal in one year), is to reduce one's use of fossil-fuelled transport. An important part of doing this is trying to ensure that your home is within walking or cycling distance of where you work and the other services that you need and use – shops, park, doctor's surgery, post office and schools for example – or a means of public transport exists which can take you to them.

The Code for Sustainable Homes

This is a voluntary standard of environmental performance introduced by the government in 2007 with the aim of encouraging house-building companies and others to build to higher standards than the mandatory minimum laid down by the Building Regulations. It is similar to and supersedes the EcoHomes standard. Licensed assessors rate homes according to measures of energy performance as well as water efficiency, sustainability of materials specification, surface water management, waste management, pollution, health and wellbeing, residents' information, contractor's environmental performance and ecological impacts. The ratings start at Level 1, which is a 25% improvement on the baseline Building Regulations energy performance, rising to Level 6 which is a zero-carbon standard, i.e. no net carbon emissions are permitted for heating, hot water, ventilation, lighting AND electrical equipment and appliances. This is a relatively high standard, which requires not only a very energy-efficient home but also sources of renewable energy – particularly renewable electricity, which can be difficult to provide. Government targets are for all new homes to be carbon neutral by 2016 (some authorities including Wales and London are aiming for this to be achieved earlier: by 2012). Anyone aspiring to build an ecohouse should aim for Level 4 or 5. All new social housing has to achieve Level 3, and a handful of private developers are moving towards Levels 4 or 5.

The Code for Sustainable Homes does not measure the proximity of the site to local amenities, unlike the EcoHomes standard, which awarded credits for having a shop within 500m (about a five-minute walk away), and other amenities or public transport links within 1km or 10 minutes away, implying a reduced use of the car. In this connection, both the Code and EcoHomes require convenient and secure storage for cycles.

The ecological value of the site

When considering how to develop a site, you should bear in mind one or two straightforward principles. Firstly, protect existing ecological features – trees, hedges, ponds and streams – as they contribute to biodiversity and mitigate extremes of climate. They can be damaged by fire, pollution from diesel or cement spilt on the ground, compaction of the soil caused by diggers driving about, and changes in the water table. They need to be protected by strong fencing, which in the case of trees needs to be positioned to include the area below the spread of the tree. Secondly, develop the low-ecological-value areas of the site. It may seem paradoxical not to develop the best parts of the site, but when you think about it there is sense in building on the bad bits and keeping the good bits. Finally, enhance the ecological value of the site with new plantings of trees, with native plants which are food plants for birds and insects, and new ponds and streams which also support wildlife. This is discussed in more detail in the chapter on Designing a Sustainable Garden (page 252).

Noise

The principal sources of excessive noise tend to be roads or aircraft, which are often a considerable nuisance but can adversely affect your health in extreme cases.

The siting and orientation of buildings in relation to the noise source should be considered in order to reduce harmful effects. Locating window and door openings away from the noise source can help, as can the grouping of buildings into courts. Also consider noise sheltering through earth banks, physical screens and walls or planted shelterbelts. Noise attenuation can vary between 1.5 to 30dB per 100 metres of shelterbelt, depending on the type of vegetation.

Once siting and forms of sheltering have been considered, think about the specification of the building itself. Noise from outside a building can enter through windows, ventilators, walls, roofs and doors. If walls or roof are constructed from lightweight materials, they may allow transmission of significant amounts of sound into the building. Ensure that all ventilators have been acoustically detailed. In most cases windows provide the main path, so it is important to ensure that their insulation is specified correctly and that they are fitted with effective seals. With double-glazing, in general the wider the spacing between the panes, the higher the insulation against sound. Secondary glazing, with an optimum distance between the glass of around 100mm (as opposed to sealed double-glazing units) provides better insulation against noise at high frequencies, such as electric trains, but may be only marginally better against low-frequency noise such as that from road traffic. Make sure that the reveals between the two layers of glass are lined with a sound-absorbing material. Thicker glass improves the performance substantially. '2+1' triple-glazing in the form of two sashes coupled together, one with sealed double-glazing and the other with a single sheet of glass, provides even better acoustic performance, as well as giving high thermal performance.

Flooding

Flood plains cover around 10% of the land in England and Wales, and nearly 6 million people live on them. The rate of development on flood plains has more than doubled in the past 50 years.

Adopt a precautionary approach: check the flood plain maps for your area (these are available from the Environment Agency website), and avoid building on a flood plain wherever possible. Apart from increasing the risk of flooding, developing flood plains may damage special habitats with rich flora and fauna; such building reduces the capacity to store flood waters, thereby increasing the risk of flooding downstream, and increases water runoff from impermeable roads and roofs, thus increasing the risk still further. The more extreme weather conditions predicted for the UK as a result of climate change may lead to more frequent flooding in some areas.

Air pollution

One fifth of humankind live in places (particularly cities) where the air is not fit to breathe. A cocktail

Left: Many of us live in cities where the air is not fit to breathe.
Right: Greening roofs helps to improve the microclimate in the city by providing shading, cooling, moisture retention and dust absorption.

of poisonous gases is given off by factories, power stations and aeroplanes, which combined with exhaust fumes from vehicles, causes respiratory diseases such as asthma. Air pollution from cities produces a cocktail of sulphur dioxide and nitrogen oxides, the byproduct of which is acid rain. Another byproduct is ground-level ozone, which has a highly corrosive effect on vegetation, leading to crop damage. Air pollution weakens trees' resistance to disease, and makes them more prone to fungal disease.

Large urban areas create particular local climates which trap pollution, have less solar radiation and are warmer than the surrounding countryside by several degrees. In summer this can lead to uncomfortable and unhealthy conditions.

Planting can be used to improve the microclimate in a number of ways:

- Use planting, water features and trees, and plant to increase cooling, shade and dust absorption. Select tree and shrub species which are tolerant of pollution.

- Green roofs are another excellent way of modifying the microclimate. They insulate the building, retain water (which reduces storm-water runoff), cool the urban heat effect, and encourage wildlife.

- Growing foliage on walls can shield buildings from extreme temperatures and frost as well as sheltering the external envelope from rain.

- Trees are particularly good as air and noise pollution filters, particularly near busy roads. Dust and pollutants adhere to dry twigs and leaves, removing 75% of dust particles.

Land contamination

Developing contaminated land affected by industrial processes and the use of toxic chemicals which are harmful to human health as well as other ecosystems is an increasingly important issue, as the current emphasis is on using brownfield sites within existing settlements in preference to previously undeveloped greenfield sites. The objective of a more sustainable form of development should be to deal with contamination in the most benign way possible.

Where contamination exists, it is often perceived as a barrier to redevelopment, and many people are unwilling to live on previously developed land. However, current planning policy means that brownfield sites will inevitably be recycled into new uses in order to protect greenfield sites. Cleaning up contaminated sites is an environmentally desirable consequence of this.

Very few sites are so badly damaged that they cannot be reused at all, but the type of contamination and the costs of clean-up will determine whether the site is suitable and financially viable for a house. When making a planning application for a site which might be contaminated, you will have to provide information on soil conditions and make proposals for dealing

Left: Greening walls also benefits the microclimate in cities as well as reducing the exposure of buildings to wind and rain.
Right: The local authority has recently let out this roof to a mobile phone company – but is there a risk to the health of the tenants?

with any contamination present. Site decontamination is a very complex subject, and beyond the scope of this book: expert guidance should be sought.

You will have to balance the cost of the decontamination process against the environmental impact of the decontamination process itself. The most benign method involving least energy is biological decontamination using toxin-neutralizing plants such as willow saplings and reeds, but this takes time and requires careful preplanning and process management. Treatment with chemicals can be quicker, but uses a more energy-intensive product. Capping the contamination with clean soil, for example, does not get rid of the problem but may be cost-effective if contamination is not severe. The last resort is scraping the site clean and exporting the contamination, which may be essential for development, but is the least environmentally desirable, as it just transports the problem elsewhere and buries it. The economic viability of this method is also affected by the cost of disposing of contaminated soils in landfill sites.

Electromagnetic fields

This section examines the potential hazard of electromagnetic fields outside the home, caused by overhead power lines, electricity substations, large transformers and mobile phone masts. There is concern, but no conclusive evidence either way, that continued exposure to magnetic fields is harmful to human health and in particular is carcinogenic. There is no officially safe level set in Britain. Electromagnetic fields within the home are discussed in the chapter on the potential risks to health (page 198).

Whenever electricity is used, an electromagnetic field (EMF) is produced. There are two types of field produced: electric fields resulting from a voltage differential, and magnetic fields arising from a flow of current. Both diminish rapidly with increasing distance from the source. Electric fields can be screened by most common building materials, whereas magnetic fields will pass through most materials.

The principal sources of electro-magnetic fields are:

- High-voltage transmission lines on pylons, which carry high current and therefore give off both high electric and high magnetic fields. Powerwatch, an independent pressure group, states that people living within 200 metres of overhead lines may be affected.

- Low-power substations are found every few hundred metres apart in a typical urban area, and many existing housing developments contain substations. Again, minimize risk by siting housing away from substations or screening using thick concrete walling or similar which can screen electric fields.

- There has been a public debate over the safety of mobile phone masts, and this will continue as the need for more masts grows with the proliferation of mobile phone use. The Stewart Report published in May 2000 gave the findings of an independent expert group set up by the UK government. It recommended that a precautionary approach be adopted when siting mobile phone masts, particularly in relation to schools and other sensitive sites. Generally, high masts (say 15 metres) are safer than lower lamppost type masts of 6.5 metres which may direct radiation straight into homes.

Costs and benefits

There are costs associated with dealing with noise, soil contamination and other site problems. Make sure that this is reflected in the valuation of the site. There is a cost in enhancing the ecology of the site, which should be reflected in the quality of the living environment you create.

Building for longevity

You should keep the future in mind when designing or commissioning a design for a house. Your needs and expectations may change; you should pay particular attention to needs that will arise if you should become infirm; and your house should be durable so that it has a long, useful life.

Adaptability

Any system needs to be capable of adapting to changing circumstances for it to survive and be sustainable. Sustainable homes need to be designed so that they are capable of adapting, which they may do by being of a simple form that can be adapted to different uses; or they may be fitted with sophisticated service systems to cater for changing demands. This section looks at some of the implications of adaptable building.

Needs and expectations change

Needs change as families grow up. Teenagers need space for homework, a place to make noise, and the ability to come and go late without disturbing other members of the household. Whilst grandparents tend not to live with their children so much in the latter part of the twentieth century, an inability of the health services to provide adequately for an ageing population may mean that more space for the older generation may be needed in the future. Employment is less secure than it has been in the recent past, so the opportunity to let accommodation to generate income or to work from home may be required. Patterns of work are changing as global corporations outsource more services and more people work from home as self-employed entrepreneurs.

One of the most significant events that can change what is needed of a house is if one of the household members becomes infirm and not able to use the bath or toilet, to walk or to go upstairs. This can happen through illness, age or accident. When planning a home, a number of simple measures can be incorporated at the outset at minimum cost which will permit the house to be used by a person with limited mobility. An example would be to make the doors sufficiently wide to accommodate a wheelchair; adapting them later is an expensive and

disruptive process. A house with level access and wide doors is also more convenient for everybody – for a young family with a push-chair for example.

Your expectations may change; what was considered an adequate kitchen 15 years ago may now need to be twice the size to accommodate a double bowl sink, the freezer, the dishwasher, the full height fridge, the high-level oven and the microwave, as well as space for the food mixer, the toaster and the coffee machine.

Our sources of fuel are going to have to change radically to address the exhaustion of fossil-fuel reserves and the more critical issue of emissions causing climate change. Conventional fuels will become more expensive as stocks become harder to exploit and as policies to reduce the use of fossil fuels begin to bite. Whilst you may design now to conserve energy and to use renewable energy sources, the technologies involved with solar power, micro-wind power, hydrogen fuel, heat pumps and combined heat and power installations are all developing rapidly. The demands made on buildings as a result of using different energy sources may be significant: for example storage volume and flues for biomass, or orientation, elevation and lack of shading for solar power. You should consider how to 'future-proof' the house and reduce the cost of adapting to other energy sources in the future.

Similarly, the technologies for television, telephone, data and security are all developing and will make different demands on our homes in the future.

A sustainable system of any kind must be capable of adapting to changing circumstances. Building houses with adaptability in mind means that they should have a long useful life before they have to be scrapped and replaced, which helps to conserve resources. Homes can also become no longer useful because of inadequate budgets and poor standards of construction. Building for durability and minimizing the need for maintenance can also reduce the need for resources.

Long-life housing

Many Georgian, Victorian, Edwardian and inter-war houses remain useful because:

- They are built to relatively generous space standards. Neither the house nor the rooms in it are built to the minimum sizes of modern homes. The additional space above the minimum allows the dwelling to accommodate changing needs.

- They are built in forms that can be extended relatively easily: terraces or semi-detached and detached houses. The use of traditional roof purlins and rafters allows extension upwards into the loft, which is particularly useful in high-density situations where extending the footprint of the house is often not possible. Modern multi-storey blocks, however, are almost impossible to extend.

- They are built in a way that can be modified: load-bearing brickwork can have openings made in it, although other constructional systems such as frame structures and dry methods of building such as timber construction are cheaper, easier and less disruptive to adapt.

- Services are capable of being upgraded relatively easily; open coal fires and gas lighting have given way to central heating and electric lighting. There are accessible voids in the floors and roof, although services generally have to be chased into walls. There are modern bathroom and kitchen unit systems that accommodate accessible services.

Separating layers with different life cycles

It is useful to think of "a properly conceived building as several layers of longevity of built components". The layers have been expressed by Stuart Brand in his book *How Buildings learn: what happens after they are built* as the six S's, which are:

- *Site*. This is eternal.

- *Structure*. The foundations and load-bearing elements are difficult and expensive to change. These are the essential, unchanging building which may last 30 to 300 years.

- *Skin*. The envelope of a building may change every 20 years or so, to keep up with improvements of thermal performance for instance, or to replace parts which have deteriorated.

- *Services*. Wiring, plumbing, ventilation and heating systems and lifts, all of which wear out or need to be upgraded at between 10- and 15-year intervals.

- *Space plan*. Interior layout, which may change frequently in offices, but at perhaps 15- to 30-year intervals in housing.

- *Stuff*. Furniture and equipment that are moving around all the time.

Buildings need to accommodate these different cycles of change for them to last well and improve over time rather than degenerate. Houses should be designed so that the short-life layers can be changed independently of the long-life structure.

Experience in other countries

In some countries an adaptable approach to building is much more common than in the UK. In the Netherlands there is a part of the house building industry that has followed the ideas of the architect N. J. Harbraken, who developed the notion of

Pipes and wires in a service void – separating the services from other layers of the building enables them to be changed, improved and replaced easily.

'supports', or long-lasting structural frameworks which are filled in with shorter-lived non-structural walls and partitions. Residents have a choice of layout and specification using prefabricated components. This combines the idea of residents participating in the housing process with the idea of economic sustainability deriving from the efficient use of resources and reduction in waste from using advanced technologies and prefabrication. In Japan, too, the house-building industry uses a high level of prefabrication, with 'Open Building' thinking which establishes a framework within which a wide range of different components can be assembled to form different houses.

Lessons from the commercial sector

Some aspects of this approach are now common practice in the construction and fitting out of commercial office space. Technologies have been developed to permit prefabricated plumbing and wiring and soundproof dry partitioning for example, and some of these techniques are being tried in the housing sector, particularly in the Netherlands. Also, the move towards loft-living is creating larger, adaptable dwelling types.

What are the engines of change?

There is a growing awareness amongst the British public of the higher standards of space and equipment enjoyed by households in other parts of Northern Europe and North America. People's expectations of space and performance are rising.

Household composition is smaller and more dynamic. Young people leave home earlier, and old people live with their offspring less often. This has to be set against the reduced ability of the health service to care for old people, and so the granny flat becomes more desirable.

Also, modern electronic systems offer many potential advantages in the home, and the demand for these benefits is likely to have a significant impact on the design and specification of houses over the next five to ten years. We will expect cabling systems for voice, data, entertainment, security, building-management systems to reduce energy consumption, remote diagnostic services for people who are old or unwell, as well as ever more demanding computer links for home working. The Joseph Rowntree Trust Smart Homes project demonstrates some of the possibilities in this area.

How much should we invest in adaptable buildings?

There is clearly additional cost in providing more space inside and space for extensions outside, for additional rooms and adaptable construction methods and accessible services. This has to be balanced against additional costs in the future of adapting housing which has not been designed with adaptability in mind to new needs and expectations. There is, however, the risk of high opportunity costs inherent in investing for future change which does not take place. We have to make decisions based on what we know now and what we can learn from the past. One or two rules of thumb could be derived:

- Do not invest too much on adaptability to meet a future that is very difficult to predict.

- On the other hand, recognize that change is inevitable.

- Plan for foreseeable change to reduce risk. This might include more home working and higher standards of space and equipment in the home.

- Simple measures that may cost little can keep future options open – non-load-bearing construction, for example.

FACING PAGE Top: The dining area of Ken Atkins' house in Lewisham, seen here over Walter Segal's head, was too small when the whole family came round. Middle: Ken was able to buy six new woodwool panels and he and a friend had the extension watertight after two weekends' work. The existing kitchen window and the patio door were fixed in a new position. Bottom: Inside the new extension which only cost around £1,500 at the time which represents amazing value.

Costs and benefits

With a house designed for adaptability, you can look forward to more space and greater convenience and reduced costs for extensions in the future. Houses that can adapt will be useful longer reducing the future use of resources. Building in adaptability has a present cost but can save money in adaptations and premature obsolescence in the future. Some common sense and relatively inexpensive measures now can save money in the future.

Summary: creating an adaptable house

Adaptability in housing is a necessary precondition of sustainability. Only if homes can adapt to changing needs and expectations will they have a long useful life. thus reducing the waste of resources implied in premature demolition. Adaptability will have to deal with a number of foreseeable trends including rising expectations of space and equipment, the increase in the number of smaller households and greater reliance on electronics in the home.

Adaptability can be achieved by:

- Generous space standards

- Adaptable built forms – terraced houses rather than apartment blocks for instance

- Providing space for expansion of dwellings

- Built-in expansion space, in the loft for example

- Easily modified structural systems such as frames

- Easily accessible service installations

- Easily adapted forms of building such as timber construction

There are a number of useful concepts, including separating parts of a building which have different replacement time scales, and in particular separating structure and infill, as is common in office buildings.

The Segal Method

This timber-frame method, devised by the architect Walter Segal in the mid-60s and used successfully for 200 or more self-build houses since built individually or in groups, is a practical example of the separation of the S's referred to by Stuart Brand. The structure stands on, but is not fixed to, the long-lasting foundations. It provides a long-term framework which is filled in with non-structural wall panels, partitions, windows and doors which can be altered to suit particular needs and wishes. The services are run in voids in the walls, floors and ceilings so that they are accessible and can be modified and renewed as necessary without damaging the fabric of the building.

Segal Method buildings demonstrate one or two general approaches which can improve the adaptability of buildings. So provide accessible service voids and build large span frame structures with non-loadbearing infill.

The method is devised to be economical and quick and relatively straightforward to construct. A Segal building is supported on individual pad foundations which consist of blocks of concrete, commonly 600mm square in plan and 900mm deep, under each post which are spaced around 3-4m apart. These foundations can be hand-dug and there is no need to level the site which eliminates the need for heavy earth-moving which makes the construction straightforward for self-builders.

The post and beam frames are bolted together flat on the ground. The level of the foundations is taken and the posts are cut to length before raising the frames into position. They are braced in position temporarily whilst the roof and floor joists are fixed. A piece of lead is placed between the bottom of the post and the foundation to prevent moisture getting into the end grain of the timber leading to rot.

The infill of readily available, standard building materials; maintenance-free, weatherproof cladding panels on the outside, a finish of plasterboard on the inside with a core originally of wood wool, which is a board no longer available made from a mix of wood strands and cement, is fixed within the post and beam timber frame using dry fixing methods such as bolts and screws. The building is laid out on a modular grid based on the stock sizes of the materials to avoid cutting and waste. The grid is a so-called tartan grid with a zone of 50mm which accommodates the thickness of the infill panels and the structure separated by a spacing of 600mm which is the width of commonly available panel materials.

The external and internal walls are not load-bearing which means that one can place openings in any position in the external walls and can position partitions anywhere you like on the different floors of the building. Importantly it also means that you can change your mind as you go along. You can stand in the part completed building and see which way has the view and which direction the sun is coming from and adapt the layout to suit. This can be very useful for self-builders who often find visualizing the inside

and outside of buildings difficult from plan, section and elevation views. Self-builders have also found the idea of the grid a useful way of visualizing spaces and the size of rooms. The opportunity to fix the position and size of openings relatively late in the building process is not an opportunity you have with other forms of construction which tend to be fixed once the walls start going up. The separation of frame and infill means that the houses can be relatively easily adapted and extended to changing needs and expectations in the future.

The house shown was extended onto the veranda to form a large dining area. The owner bought six new panels and he and a friend had the new extension weathertight after two weekends work at a cost of around £1,500 in about 1987. He was able to reuse the existing window and patio door. This is a much quicker and more economical extension than the conventional masonry construction.

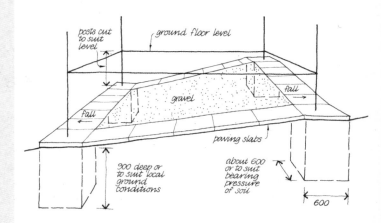

PHOTOS (left to right)
1. Foundations: individual bases under each post. 2. Making the frame: bolting together posts and beams flat on the ground. 3. Standing the frames up: adjusting for position and vertical and fixing temporary bracing. 4. Fixing the floor: the columns are the only fixed elements – walls, windows, doors and partitions can infill between the roof and the floor in any position and can be changed around in the future.
5. The finished house: the walls are made of readily available standard panels.

DIAGRAMS (above right)
Top: The foundations can be put in at any level to suit the site, eliminating the need to flatten the house plot.
Middle: The structural columns and walls and partitions are laid out on a modular grid based on the standard dimensions of the materials.

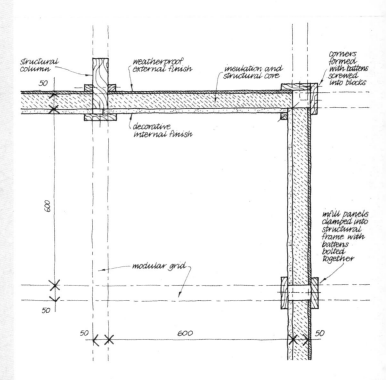

Durability

This section suggests simple approaches to increasing the durability of dwellings which can reduce the use of resources in the long term. It also deals with reducing maintenance which can be a significant cost and a substantial nuisance.

The need for durability

Along with adaptability, durability is another aspect of building houses with a long useful life. Houses are commonly designed to last 60 years, and major refurbishments are often designed to last 30 years. In practice most houses have to last much longer than this. At current rates of construction and replacement, a house built in Britain today will have to last longer than the pyramids!

Fortunately it is relatively easy to design houses to last much longer. Designing buildings so that parts which have worn out or become obsolescent can be changed easily is described under 'Adaptability' above. What has to complement this approach is the use of durable products and materials and detailing for durability. These measures can reduce the use of resources and energy needed to provide and maintain houses over their lifetime.

It is also useful to specify materials which are self-finished – which do not require applied finishes which require maintenance. Examples are durable timber used in its natural state without an applied finish, and self-coloured render.

Good maintenance is a very important part of reducing the use of resources, and is a necessary complement to adequate specification of materials and designing details for durability. Make sure that there is adequate access, and pay particular attention to services by providing plenty of access ducts and access above glazed areas such as a conservatory or lean-to roof.

Durable materials and details

Tough, weatherproof materials should be used in exposed locations on the exterior of the building. Vulnerable details should be avoided, such as exposed parapets at roof level. Details should be designed to protect the building, such as designing deep roof overhangs and setting windows back in the openings away from the face of the wall.

The durability of products and materials should be balanced against their environmental impacts. Some inherently durable materials require significant amounts of energy to produce. Life Cycle Analysis is one tool for assessing this balance, and

forms the basis of the Green Guide for Housing Specification. The green guide also gives guidance on replacement intervals and initial cost.

Costs and benefits

Durable buildings will consume less resources in the longer term, but may cost more in the first instance.

Summary: creating a durable home

The use of durable materials and details which tend to protect the building are relatively simple to implement and can extend the useful life of buildings. Durable specification and detailing has to be complemented by designing for adaptability and by good maintenance. The use of durable materials should be balanced against their environmental impact.

- Avoid vulnerable details such as exposed roof parapets.

- Design to reduce exposure, with deep roof overhangs and windows set deep within the thickness of the wall, for example.

- Specify durable materials.

Avoid using materials which are dependent on regular maintenance for their longevity – conventional painted windows for example.

Accessibility

This section outlines the advantages of Lifetime Homes standards, which seek to improve accessibility for people with limited mobility above the minimum mandatory Building Regulations standards.

The Benefits of Lifetime Homes

Lifetime Homes (see pages 146-7) offer accessibility and design features that make the home flexible enough to meet whatever comes along in life: a teenager with a broken leg, a family member with serious illness, or parents carrying in heavy shopping and dealing with a pushchair. Lifetime Homes have sixteen design features that ensure a new house or flat will meet the needs of most households. The Lifetime Homes concept, which was developed on the continent some time ago, has been adopted as standard by a number of local authorities and housing associations. It will tend to make homes have a long useful life, and thus reduce the use of resources.

Part M of the Building Regulations covers accessibility, and Lifetime Homes features add to this the built-in flexibility that makes homes easy to adapt as people's lives change.

Costs and benefits

Lifetime Homes standards offer greater accessibility and convenience for all residents including disabled people, old people, people with young children and people with short-term problems such as a broken leg.

The additional cost of Lifetime Homes Standards is marginal, with the exception of the requirement to provide space and service connections for a shower on the ground floor. The extra space required is not significant in the case of larger dwellings with a ground-floor WC. The potential extra cost is providing a ground-floor WC and space for a shower in a two-bedroom house, although a ground-floor WC could now be considered to be normal.

Summary: creating a convenient home

Consider incorporating the 16 design features of Lifetime Homes relating to the inside and external access to the dwelling.

FACING PAGE *Left: Natural oak cladding without any applied finish eliminates the need for maintenance.*
Right: Design to reduce exposure – for example by setting windows back from the face of the wall.

Lifetime Homes standards

1. Where there is car parking adjacent to the home, it should be capable of enlargement to attain 3300mm width.

The general provision for a car parking space is 2400mm width. If an additional 900mm width is not provided at the outset, there must be provision (e.g. a grass verge) for enlarging the overall width to 3300mm at a later date.

2. The distance from the car parking space to the home should be kept to a minimum and should be level or gently sloping.

It is preferable to have a level approach. However, where the topography prevents this, a maximum gradient of 1:12 is permissible on an individual slope of less than 5 metres or 1:15 if it is between 5 and 10m, and 1:20 where it is more than 10m. (Providing there are top, bottom and intermediate landings of not less than 1.2m excluding the swing of doors and gates.)

Paths should be a minimum of 900mm width.

3. The approach to all entrances should be level or gently sloping (see standard 2 above for the definition of gently sloping).

4. All entrances should:

a) be illuminated
b) have level access over the threshold and
c) have a covered main entrance.

The threshold upstand should not exceed 15mm.

5. a) Communal stairs should provide easy access, and b) where homes are reached by a lift, it should be fully wheelchair-accessible.

Minimum dimensions for communal stairs
Uniform rise not more than 170mm
Uniform going not less than 250mm
Handrails extend 300mm beyond top and bottom step
Handrail height 900mm from each nosing
Minimum dimensions for lifts
Clear landing entrances 1500x1500mm
Minimum internal dimensions 1100x1400mm
Lift controls between 900 and 1200mm from the floor and 400mm from the lift's internal front wall

6. The width of the doorways and hallways should conform to the specifications below:

Doorway clear opening width (mm)	Corridor/passageway width (mm)
750 or wider	900 (when approach is head-on)
750	1200 (when approach is not head-on)
775	1050 (when approach is not head-on)
900	900 (when approach is not head-on)

The clear opening width of the front door should be 800mm. There should be 300mm to the side of the leading edge of doors on the entrance level.

7. There should be space for turning a wheelchair in dining areas and living-rooms and adequate circulation space for wheelchair users elsewhere. A turning circle of 1500mm diameter or a 1700x1400mm ellipse is required.

8. The living-room should be at entrance level.

9. In houses of two or more storeys, there should be space on the entrance level that could be used as a convenient bed-space.

10. There should be:

a) a wheelchair-accessible entrance-level WC, with
b) drainage provision enabling a shower to be fitted in the future.

The drainage provision for a future shower should be provided in all dwellings.

Dwellings of three or more bedrooms
For dwellings with three or more bedrooms, or on one level, the WC must be fully accessible.

A wheelchair-user should be able to close the door from within the closet and achieve side transfer from a wheelchair to at least one side of the WC. There must be at least 1100mm clear space from the front of the WC bowl. The shower provision must be within the closet or adjacent to the closet (the WC could be an integral part of the bathroom in a flat or bungalow). But please note that it is important to meet the Part M dimensions specified to each side of the WC bowl in entrance-level WCs (diagrams 10a and 10b). The Lifetime Homes standards for houses of three bedrooms or more require full side transfer from at least one side of the WC.

Dwellings of two or fewer bedrooms
In small two- bedroom dwellings where the design has failed to achieve this fully accessible WC, the Part M standard WC will meet this standard

11. Walls in bathrooms and toilets should be capable of taking adaptations such as handrails.

Wall reinforcements should be located between 300 and 1500mm from the floor.

12. The design should incorporate:

a) provision for a future stair lift
b) a suitably identified space for a through-the-floor lift from the ground to the first floor, for example to a bedroom next to a bathroom.

There must be a minimum of 900mm clear distance between the stair wall (on which the lift would normally be located) and the edge of the opposite handrail / balustrade. Unobstructed 'landings' are needed at top and bottom of stairs.

13. The design should provide for a reasonable route for a potential hoist from a main bedroom to the bathroom.

Most timber trusses today are capable of taking a hoist and tracking. Technological advances in hoist design mean that a straight run is no longer a requirement.

14. The bathroom should be designed to incorporate ease of access to the bath, WC and wash-basin.

Although there is not a requirement for a turning circle in bathrooms, sufficient space should be provided so that a wheelchair user could use the bathroom.

15. Living-room window glazing should begin at 800mm or lower and windows should be easy to open / operate.

People should be able to see out of the window whilst seated. Wheelchair users should be able to operate at least one window in each room.

16. Switches, sockets, ventilation and service controls should be at a height usable by all (i.e. between 450 and 1200mm from the floor).

This applies to all rooms including the kitchen and bathroom.

Reducing energy in use

This chapter gives background information on the issues which need to be addressed to reduce the environmental impacts of the energy used to heat and light homes, and gives advice on how to reduce those impacts. The more general question of standards of performance is considered first, followed by sections on how to meet these standards, considering site layout, house planning, construction, and finally the efficient design of heating and ventilation systems. Once the base load is reduced as far as possible, the potential for renewable energy sources replacing fossil fuel is explored. Finally, the concept of offsetting any carbon emissions by planting trees which absorb CO_2 is examined. Each section has a brief statement of actions which you can take to reduce the environmental impacts of heating and lighting your house, together with a short note on their costs and benefits.

Energy Strategy

This section explains the importance of energy use, and national policies to reduce consumption and consequent emissions; outlines the various standards that are used; and suggests some principles for designing low-energy homes.

The significance of energy consumption

Reducing the use of fossil fuels and consequent emissions of pollution is generally agreed to be the most important sustainability issue. Using fossil fuels leads to global warming, climate change, acid rain and smog. Buildings consume around half the energy used in Britain, and over half of that is in the domestic sector. The largest element of domestic energy consumption in our climate is for heating. Then come hot water, cooking and lighting, and electricity for other domestic appliances.

Reducing the use of energy cuts running costs. In addition, high standards of insulation and heating improve comfort. As well as reducing energy use and environmental effects, energy conservation can enhance the value of your property. A new European Union Directive on the Energy Performance of Buildings came into effect in January 2006, and one of the measures it has introduced is the requirement for Energy Performance Certificates for all buildings. This will raise people's awareness of how much energy houses consume and how much they cost to run. It is also anticipated that energy performance will affect their value.

Energy use can be substantially cut in new houses by incorporating high levels of insulation, installing efficient heating and hot water systems, using low-energy lighting and other measures. What is more difficult is to improve the energy performance of existing buildings.

Climate change and building

What are the possible effects of climate change and global warming, and how can homes be 'future-proofed' against possible changes in weather patterns? More sustainable forms of housing should be designed to have a long life (at least 60 years), so any new buildings should be designed to cope with predicted changes in weather. The following information on global warming and climate change has been developed by the UK Climate Impacts Programme, which is part of the government's research programme into the effects of climate change:

- The flooding of buildings will become more common. River flooding already causes £100-£200m insured losses each year. Flooding leads to damage of building contents, possible contamination from sewage, and structural collapse. Some buildings could become uninsurable if they are in particularly flood-prone areas. Flooding can be avoided by careful siting of buildings, and buildings can be protected by embankments, or designed to minimize damage, e.g. by using water-resistant materials.

- Subsidence is expected to increase in buildings on clay soils, due to higher temperatures and lower summer rainfall. New buildings should include improved foundation design.

- Climate change is expected to lead to more frequent storms and increased structural damage to buildings. The level of damage could be reduced if more buildings were built to proper standards than is the case at present. Higher building standards may be needed in future, but careful design can also reduce the aerodynamic load on a building.

- Increased driving rain will occur, affecting building façades and leading to more rain-penetration around openings. Good maintenance and good workmanship would help to reduce these effects.

- More intense rainfall events could lead to drainage systems (including roof drainage, sewer systems, carriageway drainage etc) being unable to cope. Drainage systems and drainage design standards may need upgrading. Attenuation measures for dealing with surface-water runoff, e.g. through the use of sustainable urban drainage systems (SUDS), lagoons, permeable hard surfaces and green roofs should be considered.

- Climate change is expected to reduce summer rainfall, so that pressures on water resources are likely to increase. Builders should therefore aim to improve water efficiency, consider water storage, and specify drought-resistant plants.

- Higher summer temperatures could lead to a significant increase in the demand for air conditioning in buildings, and hence in higher summer energy demand. Design for more shading and natural ventilation, and to increase the thermal mass of the building.

Energy Policy in Britain

The UK government was committed to reducing the annual level of CO_2 emissions by 2010 to an amount that is 20% less than emissions were in 1990. This was in addition to our legally binding target under the 1997 Kyoto Protocol of reducing greenhouse gases as a whole by 12.5% by 2012. However, Britain is unlikely to be able to achieve the enhanced target of 20%, which has now been dropped. Meanwhile, the Intergovernmental Panel on Climate Change (IPCC) and the Energy White Paper recommend a reduction of 60% in CO_2 emissions by 2050. The *Stern Review of the Economics of Climate Change* published by the Treasury at the end of November 2006 suggested that this reduction may need to be more like 80% in order to stabilize climate change: that is to say that the globe should be producing only 20% of the current level of emissions within a period of half a lifetime – which

is a very significant change, possible only with a fundamental change in the political situation.

The Energy White Paper also spells out the implications of Britain becoming an importer of natural gas as North Sea supplies run out in the next few years. Apart from creating renewed interest in nuclear power generation as an emissions-free alternative to coal, to complement clean generation from wind and wave power, it also raises the question of the security of our energy supply.

The government has reacted by setting more rigorous targets for reducing emissions from homes. After a significant delay, they published the *Code for Sustainable Homes* in December 2006. This replaces *EcoHomes* as a currently voluntary standard for improving the energy performance and sustainability of new homes above that required by Building Regulations. It includes a minimum level of energy and water efficiency as well as standards for waste, reducing impacts on health and ecology amongst others. The Code has six levels: Level 1 requires an energy performance 10% better than the Building Regulations; Level 3 requires a 25% improvement, equates to EcoHomes' 'Very Good', and is a mandatory requirement for publicly funded homes; Level 6 is a Zero-Carbon Home i.e. zero-carbon emissions from ALL energy use in the home, including appliances and electronic equipment and not just the fixed equipment providing heating, hot water, ventilation and lighting controlled by the Building Regulations. The government has set a target for all new homes to be zero-carbon by 2016.

The government is also using the planning system to reduce emissions by amongst other things increasing residential densities and encouraging renewable energy. A number of local authorities have introduced the so-called 'Merton Rule', after the London Borough which pioneered the policy of requiring developments to meet 10% of their energy requirements from on-site renewable energy sources. The London Development Plan has more ambitious targets for renewable energy.

The Building Regulations

The energy performance of new and refurbished houses and other new and refurbished buildings is controlled by Parts L1A and L1B of the Building Regulations. Part L was revised in April 2006 with the aim of improving energy performance and paving the way for the introduction of Energy Performance Certificates as required by the EU Directive on the Energy Performance of Buildings. The new Part L sets a performance target for the whole dwelling rather than the former elemental or target U-value methods, both of which are now superseded. The assessment of the whole dwelling performance provides greater flexibility about how you achieve the required standard but increases the complexity of demonstrating compliance. Measures to extend improvements to the energy efficiency of the whole dwelling in the case of extensions or refurbishment were dropped, much to the dismay of the environmental lobby. Nevertheless, if you build an extension, change the use of a building or make significant changes such as fitting new windows or a new heating system, you will have to meet enhanced minimum standards of performance.

In addition to setting standards for the energy performance of new homes, the Regulations set standards for the energy performance of the building as a whole. In addition, they set specific standards for boiler efficiency, heating controls, insulation of pipes and hot water cylinders, minimum insulation standards for elements of construction, measures to control overheating, reducing cold-bridging at junctions between walls, roofs, floors and windows and reducing air leakage. There are specific requirements to improve the energy performance of internal and external lighting in homes, to commission systems properly and to provide operating instructions for residents for heating and hot-water systems. Also included is the performance of replacement boilers and windows in existing dwellings.

Compliance is based on a seven-step procedure:

Step 1: Demonstrate that the predicted rate of carbon emissions from the dwelling, the Dwelling Emissions Rate (DER), is not greater than the Target Emissions Rate (TER). The Target Emissions Rate is the maximum mass of CO_2 emissions, in units of kg per m^2 of floor area per year as calculated by the SAP (Standard Assessment Procedure which is a standard format for calculating the thermal performance of dwellings), that the dwelling should emit. It takes into account floor area, shape and fuel, and can be calculated either using a spreadsheet downloadable from http://projects.bre.co.uk/sap2005/ or by buying software – either way it is not an entirely straight-forward procedure and more and more designers rely on buying in the services of energy consultants to deal with compliance with Part L of the Building Regulations.

Step 2: Demonstrate that the external lighting has a maximum power of 150 W per luminaire, and either:
a. uses fluorescent or compact fluorescent lamp types and not GLS tungsten lamps or tungsten halogen lamps, or
b. automatically extinguishes when not required at night, using a presence detector or similar control.

Step 3: Demonstrate that the performance of the building fabric is no worse than the minimum acceptable U-value standards (W/m^2K) below:
a. Area-weighted average of all elements of a particular type: Wall 0.35, Floor 0.25, Roof 0.25, Windows, Rooflights & Doors 2.2.
b. Individual element value (to minimize condensation risk): Wall 0.70, Floor 0.70, Roof 0.35, Windows, Rooflights & Doors 3.3.

Step 4: Demonstrate that the boiler, where proposed, has a SEDBUK (Seasonal Efficiency of Domestic Boilers in the UK), the government database of boiler efficiency, of not less than either:

a. 86%, where a mains gas supply is available and there are no exceptional circumstances, and that the heating pumping and control systems are in accordance with basic standards, or

b. 78%, in exceptional circumstances.

Pipes and any hot-water cylinder will have to be insulated, low-energy lighting fittings provided: one per 25 sq m of floor area AND one low energy fitting per four fixed fittings; fan power for ventilation will be taken into account. Renewable-energy sources such as solar water heating and micro CHP will be taken into account.

Step 5: Demonstrate that the dwelling has appropriate passive control measures to prevent excessive solar overheating. This can be done by an appropriate combination of window size and orientation, solar protection through shading and other solar control measures, and by using thermal capacity coupled with night ventilation.

Step 6: Demonstrate that the building fabric has been constructed so that:

a. there are no gaps or significant thermal bridges in the insulation.

b. there are no significant thermal bridges at junctions between elements (wall/floor, wall/roof) and around openings (windows, doors etc.).

c. the air permeability target is achieved. The target is the value used in the calculation of the Dwelling Carbon Emissions Rate, and not the minimum acceptable standard. An air test is a mandatory requirement.

A Commissioning Certificate will be required for the heating and hot-water systems.

Step 7: Demonstrate that the necessary provisions for efficient operation and maintenance of the dwelling have been put in place by providing information including operating and maintenance instructions.

In practical terms this means that typical U-values will have to be improved from the previous standards as follows:

• Ground floors will have to have a U value of no more than 0.22 W/m²K, which can be achieved by between 55 to 110mm insulation depending on the material and the shape of the building. This compares with a previous minimum standard of 0.25 or between 45 and 90mm of insulation.

• Walls will have to have a U-value no greater than 0.3, as compared with 0.35 W/m²K previously. This can be achieved by a fully filled cavity between 90 to 115mm, as opposed to 65 to 95mm previously, or a timber-frame wall with 140mm studs as opposed to 89mm.

• Windows will need to achieve 1.8 rather than 2.2, i.e. argon-filled double-glazing with low-e glass whereas argon filled units were not necessarily required previously.

Justifying compliance of constructions which are only marginally better than the minimum standard can be complex and may require difficult and expensive remedial action if for example the airtightness does not achieve the required standard. Better by far is to aim to achieve the best standard possible within practical limitations of budget and construction. The minimum standard is not high and will not achieve the level of reductions in emissions necessary to hit the targets implied by the Stern Review and elsewhere.

Other standards

Various yardsticks of energy performance are used for different purposes. One of the most useful is the NHER (National Home Energy Rating). It is similar to the Standard Assessment Procedure (SAP) referred to in the Building Regulations, but more sophisticated, taking into account local microclimatic factors, how a home is occupied, cooking, lighting and appliances. A low-energy home would score 10 or very nearly 10 on a scale of 1 to 10.

Energy strategy

The first step is to reduce energy consumption by measures that may include:

- Site layout to provide as much shelter as possible.

- Planning homes to reduce heat loss by adopting a compact built form and incorporating lobbied entrances.

- Construction of an airtight, well-insulated building envelope.

- Systems for heating, hot water and ventilation which are efficient.

It is also important to be aware of how to operate the building and its systems in an energy-efficient manner.

Having reduced the demand for energy as far as possible in these ways, you need to consider replacing energy from fossil fuels with energy from renewable sources. This might include solar space-heating and hot water, solar electricity generation and, depending on where you are, wind power, biomass fuel and other sources of energy. For more detail, please see Renewable Energy (p168).

Residual fossil fuel use and consequent CO_2 emissions can be offset by planting and maintaining trees to absorb CO_2, creating a 'carbon neutral' development. This is only relevant if emissions have been reduced by all other ways of cutting consumption and replacing fossil fuels with energy from renewable sources. Many environmentalists do not regard 'carbon sinks' as sustainable. These issues are considered in more detail under Carbon Offset (p173).

The costs and benefits of reducing energy consumption

Reducing energy consumption means extra capital costs at the outset, for instance by increasing the amount of insulation. However, if you take a radical approach and reduce energy use very significantly, you may be able to offset the initial extra cost of insulation by eliminating the need for a central heating system. Furthermore, reducing energy use and fuel costs will save money in the longer term. There may be an opportunity cost for future-proofing your house against future climate change and alternative fuels which may be offset by savings in the future

Reducing energy use:

- cuts running costs

- increases comfort

- can enhance the value of your property

- reduces global warming, climate change, acid rain and smog.

Summary

Energy conservation is the most important issue from an environmental viewpoint. It may not be easy to improve the performance of an existing building.

- Reduce energy use by considering site layout, house planning and, most important, constructing a well-insulated and airtight building with efficient systems.

- Only then consider renewable energy from the sun or wind.

- It is prudent to consider the future when designing and building in the present. New houses should be 'future-proofed' against worsening climatic conditions – more storms, rain, flooding and also drought. Britain will shortly become a net importer of natural gas, and the security of our energy supply is threatened as a result. Build in the possibility of using other fuels.

Site layout

This section considers how site layout for shelter from the wind and orientation for solar energy can reduce energy consumption.

Microclimate

The relationship of buildings to the local microclimate can reduce the amount of energy required for heating. Shelter from the wind can reduce the amount of heat loss from air leakage, and designing to capture heat from the sun can reduce the amount of heating required. These issues apply largely to new homes, but there may be limited scope to improve shelter around existing buildings.

Shelter

High winds increase infiltration of cold air and also cool the outside surface of the house, especially when it is wet from rain; this increases heat transfer from inside to outside. In Britain generally, over half the wind comes from the south-west. Around another 15% is from the north-east, but wind from this direction is on average 5°C colder. You should design with wind from both these quarters in mind.

Site layout for shelter from wind

When laying out the site, consider:

- orientating the narrow end of the building to the prevailing wind to reduce exposure

- spacing groups of buildings around six times their height apart, to maximize the sheltering effect

- planting shelterbelts of trees about as high as the building and at a distance from the building of between one and three times the height

- courtyard layouts, L-shaped plans and walled gardens all create shelter and also pleasant sheltered external space.

Orientation for solar energy

An unimaginable wealth of energy falls on the earth's surface and provides nearly all the world's energy needs through warmth, wind, rain and plant growth. The solar energy falling on the earth in one hour is equivalent to our global annual fossil fuel use. A tiny amount of this energy goes into fossil fuel reserves, but we are using these reserves at a much faster rate than they can form. To intercept some of this energy and turn it into useful electricity, hot water or warm buildings is to use the planet's energy income rather than our dwindling fossil-fuel capital. It is estimated that an effective strategy to use solar energy can provide up to one third of the space heating and one half of the hot-water requirement for a family house. It can also provide up to 70% of the electrical power requirement.

The site layout should provide access to solar energy. The spacing of buildings should prevent overshadowing. Remember that a shadow falling on one cell of a photovoltaic array can reduce its effectiveness enormously. Shading on a solar hot-water panel, on the other hand, is much less critical. The principal elevations of buildings should be orientated within 30° of south. Deciduous trees can be planted to provide shade in summer to prevent overheating of south-facing rooms and conservatories. When the leaves drop in winter, daylight and sunlight will be admitted, although it is important not to plant too densely or too close to the building or there will be too much shading in winter.

Costs and benefits

Low-energy dwellings are likely to increase their value as energy becomes more expensive and people become more aware of energy use in houses.

However, achieving adequate solar access may produce a lower-density development with higher land- and site-development costs.

> ## Summary: designing with the site in mind to reduce energy consumption:
>
> - Lay out the site to maximize shelter from the wind.
> - Orientate buildings to maximize the opportunities for solar energy.

House Planning

Having looked at how the site is laid out, this section deals with the planning of the home itself to reduce energy consumption.

Form

The form of a house can reduce heat loss in several ways, as follows:

- Reduce the exposed external surface by avoiding detached houses and by designing as compact a form as possible. For example, most flats have only the front and back wall exposed.

- Reduce the area exposed to cold winds by having a low roof on the north-east side.

- Sheltering the building with earth can achieve a similar result.

- Planting climbing plants so that they cover the walls will tend to extend the boundary layer of warmer, less turbulent air around the building and reduce heat loss.

- Deep roof overhangs help. They also shelter the walls from rain.

Above: The low roof at the back of this cottage in Norfolk protects it from the north-east wind.
Below: Climbing plants provide shelter and reduce the cooling effect of the wind, especially when it is wet.

Porches

Unheated 'buffer' spaces between the warm inside of the house and the cold outside act as airlocks when people go in and out. This prevents cold air penetrating the house, as tends to happen with a conventional British entrance hall. A draught lobby is traditional in Scotland, and should be thought of as a minimum provision.

The general principle is to use a simple construction to absorb the main force of the wind, and then to let a properly draught-proofed inner door do its job effectively. Porches should be unheated and of simple, cheap construction. They need to be at least 2m^2 in area to allow for one door to be closed before the other is opened.

Conservatories

Until recently, a conservatory was considered an unheated space that relied on solar gain to make it usable for around nine months of the year. The construction was relatively simple, with single-glazing to provide good-value space. To reduce heat loss and air leakage, the conservatory could cover as many windows and as much wall as possible.

However, many people now regard a conservatory as a light and pleasant extension of the house, to be occupied and heated all year round; some people have even added air conditioning to deal with over-heating in summer. Used in this way, a conservatory becomes not an energy-saving measure but a large extra energy load. In these circumstances, reduce the amount of glazing by having a largely solid and insulated roof with a reduced area of high-performance double-glazing and well-insulated walls and floor.

Another approach is to design conservatories as 'sunspaces' and make them too narrow to be useful as rooms in their own right, so removing the temptation to provide heating and use them all year round.

What is crucially important is that a conservatory is not considered as a glazed (and therefore

Top: A lobby reduces heat loss, and acts as an airlock.
Middle: This conservatory is single-glazed, but has an insulated roof to reduce heat loss and overheating in summer.
Bottom: This sunspace provides passive solar heat gain.

extremely inefficient) extension to the general living space, subject to massive heat loss in winter and at night and to massive solar gain causing overheating when the sun is out in summer. On no account provide heating or air conditioning.

A conservatory on the south side of the house is often the basis of passive solar heating provision. See the section on Renewable Energy (p168).

Drying space

Provide space for drying clothes by natural means, removing the need for energy-intensive electrical tumble dryers. This might be in the form of a clothes line or rotary clothes line in a garden, or a utility room with a drying rack. Individual heat-recovery ventilators, which have a heat exchanger warming the incoming fresh air, can be useful for this.

Costs and benefits

Sunspaces and porches can save energy and also provide additional living and storage space. They can also create bright sunny space relatively cheaply. Meanwhile, compact building forms will often have a lower initial cost than more complex forms.

Summary: planning the house to save energy

Consider the form of the building, perhaps providing a porch and a space for drying clothes. Treat conservatories with extreme caution.

Construction issues

This section outlines:
- ways of reducing heat loss through the fabric of the building – insulation, airtightness and thermal bridging

- the selection of insulation materials and how to determine their appropriate thickness

- how to avoid condensation

- the standards in the Building Regulations and other higher standards proposed by the Building Research Establishment (BRE) and the Association of Environmentally Conscious Builders (AECB).

Fabric heat losses

The aim of the design of an energy-efficient building is to minimize heat losses through the building envelope – that is the roof, walls, floors and windows – and heat losses through air leakage, while at the same time maximizing heat gains from the sun. Fitting thermal insulation and draught-stripping to an existing house is one of the most cost-effective ways to save energy and reduce emissions.

Thermal insulation

The first and most important measure is adequate thermal insulation. Air-based insulation uses different materials as the matrix to encapsulate air pockets. These materials can be:

- *of organic origin* from natural vegetable matter that is both renewable and recoverable on demolition. These materials generally need little energy to produce. Examples include cork, wood fibreboards, hemp, sheep's wool and loose cellulose fibre from recycled newspaper

- *of inorganic origin* from naturally occurring minerals which are generally not renewable but plentiful. Generally this involves moderately high energy inputs and consequent emissions. Examples include mineral wool, fibreglass, vermiculite, foamed glass and aerated concrete

- *of fossil-fuel origin*, which is not renewable, with generally high energy inputs, emissions and pollution implications from the chemical industry. Examples include expanded and extruded polystyrene, foam polyurethane and foam polyisocyanurate. These materials tend to have a higher performance than organic or inorganic insulation materials. Manufacture now cannot involve ozone-depleting gases.

The environmental preference is for insulation materials from natural organic sources. Avoid fossil-fuel-based materials if possible. However, high-efficiency foam boards such as polyurethane can have a place in refurbishment where other insulation materials would be too thick.

Thickness of insulation

There is no one simple prescription for this, as it depends on the type of construction and the fuel to be used for heating, among other factors. Other issues include:

- *Cost-effectiveness*. The cost of extra insulation will reduce fuel costs – up to a point. Beyond that point, the additional insulation does not save enough fuel cost to pay for itself. The critical thickness depends on the cost of the fuel used and the cost of the insulation installed. Surprisingly, however, it is not until you get to thicknesses of around 900mm that the embodied energy in the insulation is more than the energy saved over its life. In general terms, assuming common air-based insulation materials and gas heating, the common view is that anything much above 300mm of insulation is not very cost effective.

- *Practicality of installation*. This is often the overriding consideration, especially in refurbishment.

One of the principal difficulties in reducing the emissions of existing houses is incorporating thermal insulation into existing buildings, particularly the walls. External insulation can be installed with residents in occupation, but does require scaffolding. Internal insulation reduces the size of rooms to some extent, and is very disruptive and may mean having to move out whilst it is installed. Cavity walls should be filled and lofts insulated where they exist. External insulation is expensive and changes the external appearance of the building, which may be unacceptable in some circumstances. Other issues include the need for an insulation material that will not deteriorate when wet for cavity wall insulation. This rules out organic materials such as cellulose fibre for this application. Another limitation is that some materials such as cellulose fibre need to be blown into a cavity which has to be formed in some way. Such materials are not rigid in their own right, neither are they load-bearing under a floor slab, for example.

Building Regulations standards

The Building Regulations control the overall energy performance of the building, which takes into account its size, form and type of fuel. The Regulations do not stipulate the performance of the construction except to lay down minimum values of insulation of elements.

The thermal performance of an element of construction is expressed as its U-value, a measure of the amount of heat that will pass through one square metre of the construction with a temperature difference of $1°C$ between the inside and the outside of the building, W/m^2K – the lower the better.

Minimum acceptable values of W/m^2K are:

a. An area-weighted average of all elements of a particular type: Wall 0.35, Floor 0.25, Roof 0.25, Windows, Rooflights & Doors 2.2. This is roughly

bridges can occur within an element of construction and common examples include wall ties in cavity wall construction, studs in timber construction and insulation between but not over ceiling joists in a loft. These effects can reduce the performance by 10% or as much as 50%. They can be overcome by using plastic wall ties, composite timber studs and ensuring that insulation is of a thickness that covers the ceiling joists. The effect of repeating thermal bridging is taken into account when calculating the U-value of an element of construction. Non-repeating thermal bridges can occur at the junction between elements. Common examples are around windows and where the intermediate floor or the roof meets the wall. These effects can also reduce the performance of the envelope by up to 50%. The effect of these non-repeating thermal bridges is taken into account in the SAP by assuming a default figure based on common construction. You can improve on this by using accredited good practice details or by devising a particularly efficient form of construction which significantly reduces cold bridges and providing calculations which show the actual effect of the cold bridging.

Air leakage

The Building Regulations control the unwanted infiltration of air into buildings and homes. From April 2006 homes have to be air-tested. The building is pressurised by a large fan installed in the front-door opening and the amount of air leaking to the outside can be measured. The air permeability of the building will have to be less than 10 m³/hour/m² of exposed external surface at an applied internal pressure of 50 Pascals. For small buildings, this measure is roughly similar to 10 air changes/hour (ac/hr).

This is not a tough standard. The average for all houses, including very draughty Victorian ones, is estimated to be around 14 ac/hr and for houses built since 1960 10 ac/hr. A modern house would normally achieve around 8. A reasonable standard

equivalent to 100mm of air-based insulation in the walls, and between 150 and 250mm in the roof. Windows need to be double-glazed with low-emissivity coatings to achieve the required U-value.

In addition, there is a lower standard to be applied to any particular element or part of an element with the aim of preventing condensation risks.

b. Individual element values: Wall 0.70, Floor 0.70, Roof 0.35, Windows, Rooflights & Doors 3.3.

Thermal bridges

The Regulations also control the effect of thermal bridges which are non-insulating parts of the construction allowing heat through and thus reducing the overall performance. Repeating thermal

Above: Timber around windows and at the floor level can create thermal bridges which may significantly reduce the thermal performance of the construction.

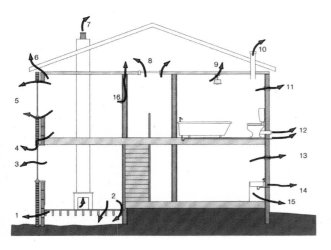

Most common air leakage paths:

1   Underfloor ventilator grilles.

2   Gaps in and around suspended timber floors.

3   Leaky windows or doors.

4   Pathways through floor / ceiling voids into cavity walls and then to the outside.

5   Gaps around windows.

6   Gaps at the ceiling-to-wall joint at the eaves.

7   Open chimneys.

8   Gaps around loft hatches.

9   Service penetrations through ceilings.

10 Vents penetrating the ceiling / roof.

11 Bathroom wall vent or extractor fan.

12 Gaps around bathroom waste.

13 Kitchen wall vent or extractor fan.

14 Gaps around kitchen waste pipes.

15. Gaps around floor-to-wall joints (particularly with timber-frame)

16 Gaps in and around electrical fittings in hollow walls.

would be 5, and a low-energy house should achieve 1 or 2 – certainly no more than around 3. Remember that the standard is based on a test conducted at a pressure difference of 50 Pascals. This is in excess of the pressure difference experienced most of the time under normal conditions, and so the actual rate of infiltration will be substantially less.

Using a smoke generator, it is very instructive to test where the draughts come from in a building. Old windows (especially sliding sash windows) and doors are very draughty. Draught-stripping old windows and doors is probably the most cost-effective energy-saving measure for an old house. Modern joinery is generally reasonably well draught-stripped, but make sure that casements are rigid enough for the ironmongery to properly close the casements against the seals, particularly on double doors. Make sure too that the frames are sealed to the structure. Other common problems occur where the first floor meets the external walls, allowing draught into the floor void which comes out at skirtings and electrical sockets and switches. Loft hatches, timber suspended floors, service entries, service penetrations into the loft and around rooflights are also common sources of air leakage. Plasterboard dry lining creates a void in the walls which can carry draught throughout the building if the external envelope is leaky.

Wet plaster on a masonry wall, or cellulose fibre insulation fully filling the cavities in a timber-frame wall, are better than dry lining or mineral wool insulation respectively; even so, they are not sufficient. You have to be positive about sealing the building at the design stage and follow through by ensuring high-quality construction on site. Sealing can be achieved either on the inside or the outside. The advantage of sealing on the outside is that you seal over the critical floor-to-wall junction. The disadvantage is that, if you have a loft, it is difficult to seal the ceiling to the outside of the wall at the eaves. This can be avoided if you design out a cold ventilated loft by having the top-floor ceiling following the underside of the sloping roof. With timber-frame construction, sealing the inside can be achieved either by bedding the plasterboard lining on mastic or by incorporating an air/vapour barrier. Sealing the outside can be carried out by a 'housewrap' of moisture-permeable building paper with taped joints. The choice of air-flow retarder is also linked to your

Airtight membranes

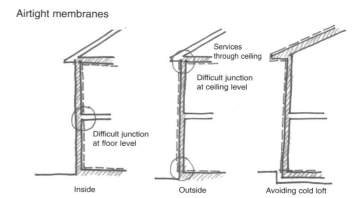

Inside   Outside   Avoiding cold loft

strategy for dealing with moisture movement and countering the risk of condensation.

The standard of construction is a critically important aspect of achieving an airtight building; contractors have to understand the need for high-quality construction and pass this on to their subcontractors, particularly the electrician and plumber, who are both used to banging holes in the structure to let pipes and wires into the building without paying much attention to sealing them up afterwards. Service penetrations often account for half the air leakage. Close supervision is required on site.

Methods which have proved effective in timber-frame construction include providing a service void for pipes and wires on the inside of the external walls so that the air-flow retarding membrane does not need to be punctured by services and requiring the timber-frame subcontractor to hand over a weathertight and TESTED airtight structure. The services are then installed and the airtightness is tested again before the plasterboard finishes are fixed.

Condensation

When warm moist air meets a cold surface condensation will take place, as on a cold window surface in winter. Condensation repeatedly running off the glass may cause the window-frame to rot. Ventilation will avoid condensation and is essential in a loft and below a suspended timber floor. If a house is well insulated and fitted with double-glazed windows, the inside surfaces will be relatively warm and there will be no surface condensation. However, moisture vapour will be driven through the construction by a higher vapour pressure inside the building towards the cold outside layers of the construction where it may condense and reduce the effectiveness of the insulation and cause decay and mould growth. This process is happening repeatedly in a masonry wall, but the materials used do not generally decay.

However, the same is not true of a timber-frame construction.

Condensation can be prevented by:

* Providing adequate ventilation to prevent the build-up of excessive moisture, providing adequate heating to make sure that surfaces are warm enough inside to prevent condensation taking place on the surface (it may still take place within the construction when it is termed interstitial condensation) and providing adequate insulation, again to raise internal surface temperatures. Minimum standards of ventilation and insulation are laid down by the Building Regulations.

* Ensuring that the construction is airtight. Much more moisture is carried into the construction through air than through vapour diffusion.

* Ensuring that the inside is resistant to vapour passing into the construction. Materials such as metal and glass are impervious to moisture; others such as a polythene sheet are very resistant. Most building materials, on the other hand, are relatively permeable to moisture vapour. Vapour resistance is usually achieved by a vapour-check layer, commonly a polythene sheet or plasterboard with a thin plastic film on the back. It is important that this layer does not have gaps which allow warm moist air into the construction. Electrical switches and sockets are a common source of discontinuities. This can be avoided by installing them in a service void inside the vapour control layer.

* Ensuring that moisture can always pass easily to the outside. In this way you make sure that moisture cannot build up within the construction. The rule of thumb is that the outside should be at least five times more permeable to vapour than the inside. In a timber-frame construction, the outside of the construction may be a fibreboard or building paper which has a ventilated cavity

behind a rainscreen cladding of tiles on a roof or render, masonry or timber on a wall. You can use relatively vapour-permeable materials and a hygroscopic insulation material, that is one that can absorb a certain amount of moisture (cellulose fibre for example, rather than fibreglass which cannot) to form a 'vapour-balanced' or 'breathing' construction. This is similar to traditional building construction (soft bricks, lime plaster and lime mortar) in contrast to most modern construction which uses hard and non-absorbent products such as glass, metal, hard bricks, Portland cement mortar and gypsum plaster. A 'breathing' construction is a risk-free solution, and some claim creates more even levels of humidity inside the building.

What is vital is that the construction must be airtight, and with high vapour resistance on the inside and low vapour resistance on the outside for condensation to be avoided.

Higher standards

The Energy Efficiency Best Practice in Housing Programme (EEBPH), sponsored by the government, publishes Good Practice, Best Practice and Advanced Practice specifications with the aim of encouraging developers to implement standards of energy efficiency in excess of minimum Building Regulations standards. See www.est.org.uk.

Good Practice represents a modest improvement, 10%, in performance over the Building Regulations and the emphasis is on improving the fabric of the building which is more difficult to improve later than the services installations which can often be retrofitted. Best Practice is a 25% improvement over the minimum and Advanced Practice is a 60% reduction in emissions and is intended for exemplar developments. The Advanced standard is related to the PassivHaus standard which has become established in Germany and northern Europe with the aim of maximizing the contribution of passive solar heating. See www.passivhaus.org.uk.

U-VALUES W/m²K					
Exposed element of construction	Estimates to achieve B Regs Part L	AECB Silver	AECB Gold	EEBPH Best Practice	EEBPH Advanced Practice
Roof	0.16	0.14	0.10	0.13	0.15
Wall	0.30	0.24	0.14	0.25	0.15
Floor	0.22	0.19	0.14	0.20	0.15
Windows	1.80	1.35	0.80	1.50	0.80
External doors	1.80	0.80	0.60	1.50 glazed 1.00 solid	0.80
Airtightness *m³/hour/m²	7.0*	3.0* for MEV** 1.5* for MVHR***	0.75*	3.0*	1.0*

** MEV = Mechanical Extract Ventilation.
*** MVHR = Mechanical Ventilation with Heat Recovery.

The AECB (www.aecb.net) has proposed a similar voluntary higher standard in response to what they see as inadequate level of performance required by the Building Regulations. This is a similar situation to Germany, where non-governmental bodies have developed low energy standards. The government there, however, now provides financial incentives for incorporating these higher standards. Other countries (Canada and Switzerland, for example) also have voluntary higher standards, but in those cases they have been promoted initially by the public sector.

The AECB Gold Standard seeks to reduce emissions further to around 10% of the latest Building Regulations standard. This is similar to the higher standards in Canada, Germany and Switzerland.

These standards also cover other energy issues including ventilation, heating equipment, lighting, appliances, thermal bridging and solar shading.

Costs

- improved thermal performance of the building envelope will increase initial costs but reduce fuel costs in the longer term

- increasing performance substantially will enable costs to be offset against a reduced cost heating system.

Summary: designing the envelope of the building to reduce energy use

- select the type and thickness of insulation in the roof, floor and walls

- consider how insulation is to be incorporated into the construction

- select the appropriate specification of windows and external doors

- avoid thermal bridges

- ensure that the construction is airtight

- avoid the risk of condensation.

Services issues

Energy consumption should be reduced as far as possible by considering shelter, form, insulation and airtightness. Then consider how the reduced energy requirement is delivered in the most efficient and least-polluting manner. This section examines fuels and heating, hot water and ventilation systems.

Fuels

Passive solar heating is preferred as it is non-polluting. See Renewable Energy (p168).

- *Bio-fuels* are good because they absorb the same amount of CO_2 as they give off when they are burnt. The most practical for domestic use is firewood, which may be a feasible choice in rural areas. Systems on a communal scale can use wood pellets or wood chips from forest waste or urban tree waste as in BedZed, a sustainable housing development in south London. Energy crops including coppiced willow and agricultural waste can also be used at this scale.

- *Fossil fuels* are the principal source of greenhouse gas and also produce other pollution including acid rain, smog, nitrous oxide and hydrocarbons. Natural gas is the cleanest alternative, followed by oil which produces half as much CO_2 again as gas, and then coal which produces two thirds as much again.

- *Electricity* is not a fuel as such, but a means of transferring energy. Apparently clean and convenient at the point of use in the home, it is in fact the dirtiest form of energy because of how it is generated. 35% is from coal (which tops the emissions table), and one third from nuclear power stations, which produce waste that cannot be disposed of or be stored safely. Renewable sources of electricity accounted for a mere 2.7% of production in 2003. Government policy is to increase this to 10% by 2010. The intricate process of converting chemical energy to heat to mechanical energy to electrical energy is only 30% efficient and thus very polluting and also expensive. It is important not to use electricity for heating and to reduce its use to purposes for which there is no convenient alternative, i.e. lighting, electrical motors in washing machines and the like, and in electronic devices such as televisions.

Electricity

Demand can be reduced by:-

- avoiding electric heating

- designing for adequate daylight

- avoiding electric cookers.

    The efficient use of electricity can be promoted by:

- fitting a low-energy lighting system with fittings dedicated to low-energy bulbs

- buying low-energy appliances; fridges, freezers, washing machines and dryers.

Nevertheless, electrical demand may still account for more than half the CO_2 emissions from a low-energy house. Of course this is all different if electrical energy is sourced from a renewable supply or if you obtain electricity from a 'green' tariff, as discussed under Renewable Energy (p168) below.

Efficient heating systems

Individual domestic gas-fired boilers can rely on natural convection and achieve up to 75% efficiency; fan-assisted boilers can achieve up to 80%; and condensing boilers can be up to 95% efficient. The 2005 revision to the Building Regulations makes a condensing boiler obligatory except under special circumstances.

This photograph of a self-build project was taken in Stockholm 25 years ago, and shows a number of features that are still only being talked about in Britain: a communal heating system and prefabricated foundation system.

A condensing boiler has an extra heat exchanger that takes heat from the hot flue gases. If the return temperature from the radiators is low, the moisture vapour created in the flue gases by combustion will condense and release the latent heat of condensation. A condensing boiler will be more expensive, but it will generally be cost-effective. It is often said that a system must be designed to ensure a low return temperature for a condensing boiler to work in its most efficient mode. This is not so, because a condensing boiler will always be more efficient than a conventional boiler because it has a far larger heat-exchanger surface area.

Combination boilers provide instantaneous hot water and remove the need for and heat losses associated with storing domestic hot water in a cylinder. This can be an efficient system and can provide unlimited hot water, unlike a storage system. Condensing combination boilers are now available, with an associated increase in efficiency. Some can serve a large household, although they will take longer to fill a bath than a storage system.

Individual central heating systems are often less than optimal in low-energy homes because normal-sized boilers and systems are often working under part load, when they are not very efficient. In a well-insulated house an individual gas convector can provide enough heat more efficiently, at lower capital cost and with less maintenance. This is combined with a small heater providing hot water in winter and summer.

A low-energy house fitted with a whole-house controlled ventilation system with heat recovery can be effectively heated by introducing a small amount of

top-up heat into the supply air when necessary. This approach is discussed under Ventilation (p167).

An alternative approach is to provide a communal heating system that supplies a development from an efficient central boiler plant. This can reduce maintenance, and the plant can be designed to adapt to different fuels to suit changing circumstances. The disadvantages are that the development must be laid out in a way that permits the heating mains to each home to be routed in an economical manner, and heating will be either metered, which is expensive to install and manage, or charged on a flat rate, which can encourage wasteful behaviour and be unfair on individuals. In Britain, communal systems have tended to be installed on the cheap and have quickly become obsolete. Wider-scale district heating installations serving whole towns have been successfully installed in Europe, using waste heat from the incineration of waste from industrial processes and power stations.

Another approach is to provide a Combined Heat and Power (CHP) system. This has an engine generating electricity and uses the resultant waste heat for heating and hot water. It requires a relatively constant demand for heat and power to operate efficiently. This can be achieved by the diversity of demand from a large number of homes or by a mix of uses requiring heat and power at different times. Small-scale units are under development.

A heat pump will provide heat taken from the air or from a body of water – or more commonly from the soil (see Renewable Energy, p168).

Heat emitters

Most central heating systems rely on radiators – a misleading name because they actually deliver heat mostly by convection. However, radiant heating can achieve comfort with lower air temperatures and less expenditure of energy. The radiant component of radiators can be boosted by increasing their area as much as possible for a given output, using single rather than double panels and simple panels rather than radiators with fins. Larger radiators have to be integrated into room layouts (which may be difficult in small houses) and have to be balanced against the need to reduce the volume of water in the system. Pipework and radiators should not be oversized.

Underfloor heating has a much higher radiant component and can provide comfort conditions at lower air temperatures and thus more efficiently. Underfloor heating is more expensive to install but runs effectively at low temperatures, so solar heating systems often use underfloor heating rather than radiators.

Controls

Good controls are important to efficiency. The need is to control time and temperature for heating and hot water. Thermostatic radiator valves (TRVs) can control the temperature of individual rooms, but a room thermostat is needed in the system to prevent the boiler pumping heat around a by-pass loop when all the TRVs are closed. The Building Regulations require a room thermostat and a cylinder thermostat which turn the boiler off when there is no demand for heat. Electronic controls offer closer control than mechanical thermostats, but for maximum effectiveness they need to be easy for people to understand and use. Systems can be zoned, with separate controls for the bedrooms and work area. Efficiency is improved by a controller that optimizes the start time of the boiler by taking into account the current external climate and rate of heat loss from the house. Also more efficient is a programmable thermostat that can set different temperatures at different times of the day to match different levels of activity – higher in the evening when people are inactive in front of the television, say. Controllers that combine these functions are available.

Hot water

A low-energy house may require more energy to heat domestic hot water than for space heating. Any discussion of heating cannot be separated from the need for hot water.

Up to half of the hot-water use of a typical household can be met by solar hot-water heating.

First, reduce the need for hot water as far as possible. Nearly half the typical usage in a four-person household is for the bath, so significant reductions can be achieved by installing a shower – a low-water-use atomizing shower rather than a high-water-use 'power shower' – and also providing low-water-use washing machines (see Water Conservation, p208).

Next, reduce heat losses from the system. Do not store water at a higher temperature than necessary. 55°C is sufficient for a bath and is hot enough to deal with any risk from legionella. To reduce waste, cut the length of the 'dead leg' of pipe between the storage cylinder and the kitchen sink and washbasins as far as possible. This is less critical for large volume or less frequent uses such as the bath. Keep the pipework between the boiler and the cylinder as short as possible and insulate it. Insulate the cylinder well.

Mains pressure systems avoid the need for a frost-protected storage tank in the loft. They are also simple to install and provide good balanced hot- and cold-water pressure for showers.

Ventilation

Ventilation is required to remove smells, smoke, carbon dioxide and moisture vapour. The consequent heat loss can account for up to 50% of the heat loss in a low-energy house.

The simplest system is a 'trickle' vent, often in the form of a slot in the head of the windows. This provides background ventilation in each habitable room – living-room, dining-room and bedroom – and can be used together with a mechanical extractor fan in both the kitchen and bathroom. The fans can be switched on manually, or activated when the humidity rises to a pre-set level. Fans with heat recovery are available, which use heat from the exhaust air to warm the incoming air in a heat exchanger. The fans

The principal source of heat loss in a well insulated airtight home is ventilation – this heat recovery system extracts up to 90% of the heat from stale air and uses it to pre-heat incoming fresh air.

require both maintenance and energy to run.

Passive Stack Ventilation (PSV) avoids these disadvantages by relying on the fact that warm, stale air rises. Large diameter pipes allow this air to rise from the kitchen and bathroom and be replaced by fresh air drawn in through the other habitable rooms. Humidity-controlled inlet and extract grilles boost the rate of extraction if there is steam from cooking or bathing, and if there is humidity because a room is occupied. There is a low level of background ventilation 24 hours a day, which provides good internal air quality. The system is passive and does not rely on fans or other mechanical devices and so there is no power required and little or no maintenance needed.

More control is provided by a whole-house mechanical ventilation system with heat recovery (MVHR). This extracts stale air from the kitchen and bathroom and provides pre-heated fresh air to habitable rooms. It is a fully controlled system, and efficiencies of heat reclamation are around 70%-90%. However, it is expensive, and you should take care to use a system with large-diameter ducts and low-loss heat exchangers so that the fan power is not greater than the energy saved. Also, the envelope of the house needs to be very airtight for the system to be effective. You should aim for an air leakage rate of 1 m³/hr/m² and certainly no more than 3 m³/hr/m². Space heating can be provided by warming the incoming air. Total fan power should not exceed 1 W per litre per second of extracted air. Systems which use low-voltage direct current fans are the most efficient.

In addition, the Building Regulations require windows in habitable rooms to be openable to provide plenty of fresh air for 'purge' ventilation.

Costs and benefits

Efficient systems will reduce emissions, may have a higher initial cost but will save cost in the long term.

Summary: saving energy by specifying efficient systems

- choose a fuel that minimises environmental impacts
- reduce electricity consumption as far as possible
- specify efficient heating, hot-water and ventilation systems.

Renewable energy

Renewable energy means clean energy that is not based on fossil-fuel use with its attendant pollution. When you have reduced energy consumption as far as possible by considering shelter, form, insulation and airtightness, and have efficient heating and ventilation systems, then renewable energy can supply all or part of the remaining energy needed.

The development of and investment in renewable energy is being driven not by a fuel crisis – fossil fuels have never been so available and so cheap – but by the realization that the planet can no longer absorb the effects of pollution.

Solar heating

All buildings benefit to a greater or lesser extent from solar energy falling on the exterior and coming in the windows to warm the building. The Greeks and the Romans were aware of this PASSIVE solar heating. ACTIVE solar heating involves separating the collector (often a solar panel) from the heat store (possibly a tank of hot water) with a system of air or water to transfer heat from one to the other. The fuel crisis in the 1970s led to renewed interest in both types of solar heating, and to many new developments in North America and Europe.

Passive solar space heating

There are many possible arrangements of solar space-heating devices. Passive systems include DIRECT passive heating, when the energy is absorbed by the walls and floor of the living space and emitted from them in an uncontrolled manner. An INDIRECT arrangement places a heat store between the solar collector and the living space to moderate the range of temperature experienced. An ISOLATED arrangement separates the collector from the habitable space with insulation to reduce heat loss at night. This generally relies on air to transfer heat into the living space. Solar collectors can be windows, walls, conservatories, roofs or panels.

Windows and walls

At its simplest, face the building south, make over half the area of the south-facing wall of high-performance double-glazed windows, and arrange for a concrete floor with dark ceramic tiling and

South-facing glazing for passive solar gain: with this extent of glass there is a risk of overheating in summer, which can be limited by roof overhangs, shading and planting. Blinds to control overlooking can negate solar-heat gains.

solid partitions to absorb the heat. The risk of overheating in summer can be avoided by a deep roof overhang and planting for shading. Lay out the house to avoid overlooking, because if you need to use net curtains for privacy you will negate the object of the exercise. Heavy curtains or shutters are needed at night or you will lose all the heat gained during the day, and a carpet should not be laid because it will prevent the floor from storing heat.

Conservatories

Conservatories are popular, largely because they also provide useful space which can be comfortable without heating for nine months of the year. They act as a buffer between the inside and the outside and can pre-heat ventilation air, provide a lobby to the outside and also conduct heat into the house. Good ventilation at high and low level, and shading, are vital to prevent overheating in summer. A solid wall between the conservatory and the house adds thermal storage but heat losses are high at night. Under no circumstances be tempted to provide heating in a conservatory as you will be wasting large amounts of energy.

Roofs

Using the roof to capture energy is good when the walls are overshadowed (in a city centre for instance), but overheating is difficult to prevent. One way is to create a solar loft with a double-glazed roof and insulation at ceiling level. This will get very hot, and a thermostatically controlled fan will draw warm air into the living space. Shading and ventilation will be needed in summer, which can become complex. You can also provide a collector for solar hot water. The Netherspring self-build project in Sheffield incorporates many of these features.

Active solar space-heating systems

Active systems incorporating large heat storage capacity, which can store heat from the summer to provide warmth in winter, are needed to make a significant contribution or even supply all the heat required for a well-insulated building. The heat store can be rocks heated by warm air, or a heavily insulated water tank, but either way it needs to be big. Such systems were designed, built and tested in Britain, Northern Europe and North America in the 1970s, but have not proved cost-effective.

Solar water-heating

This was common in the United States before 1940 and was rediscovered there in the 1970s. Systems are now common in Spain, Greece, Israel, Australia and other places with plenty of sun and few fossil-fuel resources. In Britain it is estimated that half the housing stock could easily be fitted with solar hot-water panels and that they would supply around 40% of the demand for hot water.

Systems consist of a collector and a hot water store. Collectors can be flat plates or evacuated tubes. Evacuated tubes rely on a vacuum for insulation and are more efficient, collecting a useful amount of radiation even on a cloudy day. The effi-

Most new and half existing homes can accommodate solar hot water panels, which will provide up to half the hot water requirement.

ciency of flat-plate collectors can be improved by having double-glazing and a selective surface, which is a microscopic film applied to the collector plate to increase its ability to absorb solar radiation.

The collector can be mounted at low level in the garden or against a wall, or it can be out of harm's way on the roof or incorporated into the roof finish, but if it is mounted at high level you will need a pump and controller. The system will need frost protection. It can be a direct system with a frost-resistant collector and plastic pipework or, more usually, an indirect system with a separate water circuit from the collector to the storage tank, which has antifreeze in it.

The storage tank can be a conventional hot-water cylinder with an additional solar coil at low level or it can be integrated into the collector outside. This is a much simpler arrangement but it does require frost protection, using either a vacuum or expensive translucent insulation.

The optimum tilt is 32^0 facing slightly west of due south, although a collector set anywhere between SE and SW at a tilt of between 10^0 and 60^0 will perform at 90% of its optimum performance. For a family house it will need to be around $5m^2$ if it is a simple single-glazed matt-black collector, $4m^2$ with a selective surface and $3m^2$ if it is an evacuated-tube system.

In 2005, the cost of a domestic hot-water system varied between £2,500 and £4,000, and will probably be cost-effective. To be really cost-effective, you need a very inexpensive, simple collector made out of second-hand central-heating radiators and installed on a DIY basis.

Solar electricity

So much for solar energy captured for warmth and hot water. Solar energy can also be turned into electricity in a photovoltaic (PV) panel, which converts solar energy into electrical energy using silicon cells. These PV panels are often used to power calculators, which require very little power, but the cost of higher-power applications has until recently mostly limited their use to supplying telecomms installations in remote locations that cannot be supplied from the grid. The price of PV panels has fallen, and is expected to fall further as the demand for panels leads to mass production and cost reductions. The British government has followed a number of European governments and the European Union who have subsidized demonstration installations for some years to develop the technology and the market. However, unless you can obtain subsidy, PV electricity generation currently remains much more expensive than taking power from the grid. It is often used to complement wind generation to provide electricity in summer when there is less wind.

Wind power

In the UK we have a large potential wind resource. Although we have 40% of Europe's total wind energy resource, it remains largely untapped, currently meeting only 0.5% of our electricity requirements.

Individual turbines vary in size and power output, ranging from a few hundred watts as used on yachts and caravans through 5 or 6 kW machines around 1.5 to 2m in diameter suitable for a single dwelling, to 2-3 megawatts for community-owned turbines or wind farms supplying electricity to the national grid.

The power generated by a wind turbine is a function of the cube of the wind speed: doubling the wind speed increases the power output eight times. So wind speed is critical, and depends on which part of the country you are in: check the British Wind Energy website www.bwea.com to find the wind speed in your area. Height is also critical: a rule of thumb is that a turbine should be a minimum of 10m above the roof or any other obstruction within 100m such as buildings or trees, to reduce turbulence (which reduces efficiency and causes undue wear).

A wind turbine may be opposed by your neighbours or the local planning department for reasons of visual impact, noise and conservation issues. On the other hand, a number of planning authorities including Merton in South London are now requiring 10% of the energy consumed by a development to be generated from on-site renewable sources, which might include wind turbines.

Small-scale wind power is particularly suitable for remote locations that aren't connected to the national grid, where conventional methods of energy supply are expensive or impractical. Most small wind turbines generate direct current (DC) electricity. Off-grid systems require battery storage and an inverter to convert DC electricity to AC (alternating current, as used for mains electricity). A controller is also required to ensure the batteries are not over- or under-charged. It is common to combine this system with a diesel generator for use during periods of low wind speeds. Alternatively, another renewable energy technology such as solar PV could be used. Wind systems can also be connected to the national grid. No battery is required, and any excess electricity can be sold to the national grid.

Urban areas typically experience wind speeds of between 3 and 5 m/s, and a turbine between 1.5 and 2m in diameter will provide something around 10% of the average power use of 4700 kWhr/year. Such a machine will cost somewhere around £10,000 including installation, mast and controls. However, this expenditure in a rural area with good wind speeds could provide up to 70% of your electricity.

Turbines can have a life of up to 20 years, but will require service checks every few years. For battery storage systems, typical battery life is around 6-10 years, depending on the type, so batteries will probably have to be replaced at some point in the system's life. Micro-wind generation is unlikely to provide a significant amount of electricity in the short term, as this would require greater reliability, simplification of grid connection procedures and lower costs from a larger market. Meanwhile, work is being carried out to improve efficiency by using baffles to increase wind speeds over buildings.

Ground source heat pumps

A heat pump extracts heat at a low temperature from the ground, and uses the same principle as a refrigerator to cool the ground and deliver hot water

Top: Small-scale wind turbines can be effective in windy rural areas, but may never generate a significant amount of power in urban areas. Bottom: Photovoltaic panels are currently very expensive; however, you may be able to get a grant.

for space heating. The water is at a relatively low temperature and so larger radiators or underfloor heating are required. A secondary heat source is required to raise the temperature of water so that it can be used as the domestic hot water supply.

A heat pump can supply three to four times the electrical energy input as heat. However, this may not necessarily reduce emissions if the electrical input is from fossil-fuel sources. A heat pump also relies on ozone depleting refrigerants which pose environmental dangers on disposal. At 2005 costs, a heat pump will cost around £4,000-£7,000 to install for a well insulated small house, will have a substantial payback period compared with oil and bottled gas fuels, and is more expensive to run than gas heating.

'Green' electricity tariffs

The regional electricity companies (London Electricity, Manweb etc) have lost their monopoly on electricity supply so you can now choose to buy it from a company that generates from renewable sources. These include hydroelectric and wind power (methane from landfill is not strictly renewable.) It seems likely that demand for power from clean, green sources will outstrip supply in the short to medium term, although in the long term Britain is better placed than anywhere in Europe to benefit from wind and wave power. Other potential sources of renewable power include biomass fuel such as coppiced willow, and geothermal energy (warm water from deep holes in the ground, including a few disused mineshafts). The difficulty is matching demand to the fluctuating availability of renewable energy. It remains to be seen what the long-term future holds, but it is well worthwhile buying green electricity.

Cost implications

- **A simple conservatory will reduce heat loss, particularly if it pre-heats ventilation air and provides useful extra space cost-effectively.**

- More sophisticated solar heating arrangements are often not cost-effective for a well-insulated house with low demand for heating.

- Solar hot-water heating will probably be cost-effective, and will be very cost-effective if installed on a DIY basis.

- PV panels are not cost-effective at the time of writing unless grant-aided under a government or EU scheme.

- Other types of renewable energy such as wind or hydro power are not cost-effective for small-scale installations unless serving a location far from the national grid.

- Spending on renewable energy may not be cost-effective, but may be justified when all other measures to conserve energy have been taken, with a view to creating a zero CO_2 development. However it is important to invest at the outset in measures that are difficult to incorporate later – for example, putting in underfloor insulation in advance of photovoltaic panels which can be bolted on at any time when the policy and financial climate are favourable.

- Buying renewable energy from the national grid is both cost-effective and stimulates the market for clean power.

Summary

- **Design simple measures such as south-facing windows to gain benefit from solar energy**

- **Only consider more sophisticated renewable energy sources when you have taken all measures to reduce energy consumption such as increasing the efficiency of the building envelope and heating and ventilation systems**

- **Consider buying your electricity on a 'green' tariff.**

Carbon offsetting

This section outlines how planting trees can arguably offset the production of CO_2. It outlines some of the problems and how to go about establishing a 'carbon-neutral' development.

Carbon offsetting

Plants absorb CO_2 from the atmosphere and release oxygen. However, they also release CO_2 through transpiration. The important thing is the balance between these two processes. When young plants are growing, the balance is always in favour of absorbing CO_2, so when a forest is young it is putting on bulk and absorbing a large quantity of CO_2. One can calculate the amount of CO_2 produced by a development, and plant trees to offset it.

Tree-planting is not the solution to global warming, nor is it a replacement for reducing emissions at source. Reductions and switching to less carbon-intensive renewable energy together provide a way of dealing with climate change. Tree planting is no bad thing, but should not be a substitute for these measures.

The problems

Concerns have been expressed by several environmental organizations about the use of forestry as a solution to climate change.

- They are particularly concerned that some nations, notably the United States, are using carbon sinks as a way of avoiding any commitments under the Kyoto protocol on climate change instead of reducing their emissions of greenhouse gases.

- There are also difficulties in assessing the greenhouse-gas benefits of forestry projects because the calculations have many uncertainties attached.

- Even assuming that calculations about carbon uptake of a new forest over the next 10 years could

be agreed, what happens in 30 or 50 years when new woods begin to reach maturity? At least some of the wood should be harvested, potentially putting carbon back into the atmosphere.

- Another problem is the type of tree and forest planted. The best trees from the carbon point of view – fast-growing varieties that create bulk, such as eucalyptus or genetically modified poplar – are not going to create the sort of natural forests that are best for biodiversity.

- It is just not practical to rely on carbon sinks as a method of offsetting carbon emissions; to offset emissions from the UK would require an area about the size of Devon to be planted every year.

Carbon-neutral development

There are organizations which calculate the area of forest needed to offset any particular activity; they will also plant and manage the trees for a fee to achieve a carbon-neutral development. Many are sensitive to the environment, avoiding commercial afforestation; they rely instead on natural woodland restoration, community forest developments, new amenity woodland and other projects that enhance biodiversity, and social benefits such as supporting the local economy in LDCs.

Costs and benefits

A carbon-neutral development does not contribute to climate change, but there is a cost in planting and managing trees planted to offset CO_2 emissions.

Summary: offsetting carbon emissions by planting trees

After applying measures to reduce energy use and reducing the remaining reliance on fossil fuels by using renewable energy sources, you can offset any residual CO_2 emissions by planting trees – but there is debate as to how sustainable this approach is.

Reducing environmental impacts

Having looked at how to reduce the environmental impacts of the use of energy for heating and lighting your home, the next issues are to do with reducing the environmental impacts imposed by the construction of the home itself. These include minimising damage to the ozone layer; specifying timber and other materials to reduce the use of non-renewable resources and fossil fuels in the manufacture of materials, products and components; and also ensuring that your boiler produces a minimum of polluting nitrous oxide as a by-product of combustion. The role of reclaimed and recycled materials is outlined.

Materials Specification

This section outlines the significance of the energy 'embodied' in the manufacture and transport of materials. It introduces Life Cycle Analysis (LCA) and in particular outlines the method on which the Green Guide to Housing Specification published by the Building Research Establishment and incorporated into EcoHomes is based. Other useful references are given and the particular environmental implications of common building materials are outlined.

Embodied energy

The pollution caused by burning fossil fuels is generally recognized to be the most important threat to our environment. About half the energy use in Britain goes on heating and lighting buildings, and of this over half is used in the domestic sector. 30% of all the energy used in Britain is in the domestic sector. Much work has been done on energy conservation, and this is dealt with under 'Energy' below. But meanwhile, there is also a significant amount of energy used in producing the houses themselves; producing the materials and transporting them and in the construction process itself. This 'Embodied Energy' accounts for a further 8% of energy consumption. The proportion of total energy use which is this embodied energy becomes more significant as buildings are designed to be more energy-efficient in use.

It is possible to reduce the embodied energy content of a building by up to half by adopting some simple measures. Use:

Life Cycle Analysis of a building element (windows in this example) in The Green Guide to Housing Specification, *showing the ratings for each environmental impact which go to make up the summary rating for a variety of typical types of window. The chart also gives an indication of relative costs and ratings for recycled content and potential for recycling on demolition. The pie chart illustrates the proportion of environmental impacts of a typical house contributed by the window element of the construction.*

Windows	Summary Rating	Climate change	Fossil fuel depletion	Ozone depletion	Freight transport	Human toxicity	Waste disposal	Water extraction	Aid deposition	Ecotoxicity	Eutrophication	Summer smog	Minerals extraction	Cost	Typical replacement interval	Recycled input	Recyclability	Currently recycled	Energy saved by recycling
PVC-U frame, double-glazed	C	B	C	C	A	B	A	A	B	C	A	A	B	£150-£530	25	C	C	C	C
Pre-treated softwood frame, double-glazed, painted inside and out	A	A	A	A	A	A	A	A	A	A	A	C	A	£70-£320	25	C	A	B	A
Durable-hardwood frame, double-glazed, painted inside and out	B	A	A	A	C	A	A	A	A	B	B	A	A	£130-£355	35	C	A	A	A
Powder-coated aluminium frame, double-glazed	C	C	C	A	A	C	A	C	C	A	A	A	A	£270-£360	30	A	A	A	A
Aluminium-faced timber composite frame, double-glazed, painted inside	C	C	C	A	A	C	A	C	C	A	A	B	A	£275-£370	35	A	A	A	A
Glass block window	C	A	C	A	A	C	C	A	C	C	C	A	C	£185-£420	25	C	A	B	A

- Second-hand materials when possible, and materials with a high recycled content in preference to new materials

- Minimally processed materials rather than highly manufactured products

- Locally produced materials in preference to materials from far away or imported products.

Other environmental impacts

The above rule of thumb is a start, but does not take into account other environmental impacts of the use of building materials. The most commonly recognized approach to assessing materials is Life Cycle Analysis (LCA). This, as its name suggests, tracks the impact of materials during extraction, production, transport, construction, maintenance, repair, replacement and ultimate disposal.

The LCA which forms the basis of the Green Guide to Housing Specification published by the BRE analyses the following environmental impacts. The relative weightings used to build up overall ratings are also given:

- *Climate Change (36.0%)* – the global warming potential of the emissions arising from the use of materials

- *Fossil fuel depletion (11.4%)* – coal, oil and gas consumption

- *Ozone depletion (7.7%)* – gases which destroy the ozone layer

- *Freight transport (7.4%)* – distance and mass of freight moved

- *Human toxicity (6.7%)* – emissions of toxic substances such as heavy metals assessed against air quality guidelines and acceptable daily intake for humans.

- *Waste disposal (5.8%)* – material sent to landfill or incineration

- *Water extraction (5.1%)* – mains, surface and ground-water consumption

- *Acid deposition (4.8%)* – measurements of acid rain potential caused by emissions of ammonia, hydrochloric acid, hydrogen fluoride and nitrous and sulphur oxides.

- *Water pollution (4.1%)* – caused by nitrates and phosphates over-enriching streams and lakes leading to them being taken over by algae which reduces the amount of oxygen so that flora and fauna cannot survive.

- *Ecological toxicity (4.1%)* – the effect of emissions of toxic substances on the ecosystem.

- *Smog (3.6%)* – emissions of nitrous oxides and Volatile Organic Compounds (VOCs) such as solvents cause smog and ozone which cause respiratory problems.

- *Minerals extraction (3.3%)* – relates largely to the local effects of noise and dust from quarrying and mining.

The effects of embodied energy depend on the source of energy, whether burning gas or using electricity, for instance. For this reason the BRE model traces the effects of energy use including emissions of greenhouse gases, fossil-fuel depletion, smog and acid rain, rather than using the measure of embodied energy itself.

The Green Guide to Housing Specification analyses ten principal elements of a house – the roof, external walls etc. – which together contribute around 90% of the overall environmental impacts of a house over a 60-year life.

The Green Guide to Housing Specification aims to present complex Life Cycle Analysis information in a form which is easy to use by basing it on typical constructions for each building element. This has the limitation that it is not possible to disaggregate the information if you propose a similar construction but with a different insulation material, for example.

Timber Preservatives	Health	Environment	Fire	Disposal	Other	Banned (in UK)
Active ingredients						
Creosote	●	●	·		●	
Boron	·					
Arsenic	●	●		●	●	
Chromium salts	●	●		●		
Copper salts	·			●		
Fluorides	●					
Dieldrin	●	·	●	●	●	X
PCP	●	●	●	●	●	
Lindane	●	●	●	●	●	
Dichofluanid	·	●	●	●	●	
Tributyl tin oxide	●	●				
Permethrin	●	●	●			
Cu & Zn naphthenates	●					
Acypectas zinc	●					
Carriers:						
Water-based					·	
Solvent-based	●	●	●		●	

Typical chart from the two volumes of The Green Building Handbook showing in simple graphic form the environmental impacts of different building materials, in this case, timber preservatives.

Other guidance

Two other publications will be essential reference if you require information on individual materials rather than complete elements of construction. The first is the *Green Building Handbook*. This analyses environmental assessment data on the specific building materials themselves grouped under headings such as roofing materials, and presents it in easy-to-read summary diagrams.

The second reference is the *Handbook of Sustainable Building*. This is based on the Environmental Preference Method developed in the Netherlands. This offers three alternative materials in order of preference from an environmental point of view. It also lists materials which are 'not recommended' although they are often the common solution: the use of PVC for above-ground drainage, for example. The first preference is based on a practical solution with low environmental impact and no or negligible extra cost, which does not give rise to practical problems. Note that the first preference may not necessarily be the solution with the least environmental impact, which may incur extra cost or not be proven in practice. Some of the solutions are based on Dutch building practice, but this work is the basis of the advice given in Chapter 8 on designing to reduce environmental impacts, amended to accord with UK practice.

Ozone-Depletion Potential

CFCs (chlorofluorocarbons) and HCFCs (hydrochlorofluorocarbons) and other stable gases containing chlorine or bromine in their chemical structure have Ozone-Depletion Potential (ODP) and cause damage to the ozone layer. This exposes living organisms to harmful ultraviolet (UV) radiation. They also have significant Global Warming Potential (GWP), which is defined as the potential for damage that a chemical has relative to one unit of carbon dioxide, the primary greenhouse gas.

CFCs and HCFCs have been used extensively in building as foaming agents in foam plastic insulation materials and as refrigerants and fire-extinguishing gases in commercial buildings. Production of CFCs has now ceased. The use of ozone-depleting blowing agents for insulation has also ceased, and the European Union is committed to phasing out HCFC production. Note, however, that whilst all insulation materials, including foamed plastics, now have Zero Ozone-Depletion Potential, some insulation materials, polyurethane and polyisocyanurate in particular, are not inherently low-GWP materials, and have a GWP of more than 5 – which is the Standard adopted by EcoHomes to qualify for a credit. You should avoid using these insulation materials. Some notes on the environmental effects of common building materials follow.

Stone, Concrete, Brick and Glass

The pollution generated per kilogram by these materials is generally very low. However, they are used in large quantities which leads to the environmental problems associated with them: energy used in transporting raw materials and products, and the damage to the landscape and environment arising from extraction and disposal of materials on demolition.

- Recycling brick, stone and concrete for aggregates reduces environmental damage from extraction and waste disposal. Currently the maximum proportion of recycled aggregates possible whilst still achieving the technical specification for concrete is around 20%.

- Glass requires a large amount of energy to produce with the associated pollution generated. However, it has the property of trapping passive solar energy and will often show a positive energy balance. Glass can be successfully recycled, but not for the production of new sheet-glass.

- The production of fibreglass has similar implications to the manufacture of glass, although the fibres of glass are bound with synthetic resins. The release of phenols, formaldehyde and ammo-

nia from the production and hardening of the resin is controlled in the factory which reduces the environmental impact. Glass fibres are not thought to increase the risk of cancer but they do irritate the skin, eyes and mucous membranes. When working with glass fibre it is important to provide good ventilation and wear protective clothing and a mask. Measures should be taken to prevent fibres from insulation becoming airborne and causing irritation (in North America, fibre insulation is commonly supplied encapsulated in polythene). Provide boarding to create an access deck in an insulated loft to avoid disturbing the insulation and also seal penetrations where pipes and wires pass through the ceiling. Vacuum your completed house to ensure that there are no loose fibres on occupation.

- The environmental impacts of producing mineral wool are similar to those for fibreglass.

- The production of cement is a very energy-intensive process, and is now one of the principal global sources of greenhouse gases. Many cement factories use refuse-derived fuels which, although using waste as a source of energy, do pose a potential environmental threat because of the production of dioxins and other pollutants. Portland cement is an almost universal product which permits the construction of concrete and high-strength load-bearing brickwork. It does have its limitations, however. It is susceptible to cracking caused by ground movement, thermal movement and shrinkage. The use of cement mortar for brickwork prevents the bricks being reused on demolition. In many circumstances lime mortars are much more preferable – for render and mortar for brickwork. Lime requires less energy to produce because it is fired at a much lower temperature, and the hydraulic lime suitable for use in building applications also reabsorbs during the curing process the equivalent of around 70% of the CO_2 arising from its manufacture.

Metals

The extraction of metals from ores is damaging to the environment, and harmful substances can be released. Refining metals requires large amounts of energy but metals are relatively easy to recycle and it is generally economically attractive.

- Aluminium requires vast amounts of electrical energy to extract, but can be recycled to produce high-grade materials.

- Steel production from coal and iron ore causes considerable pollution. However, the energy required compared with other metals is low. Steel can be recycled, but to a lesser extent than aluminium. Steel needs coating to protect it from the weather, and zinc coating in particular may cause environmental problems. Corrosion can be avoided by adding chromium and nickel to produce stainless steel. This process can, however, lead to emissions of these heavy metals which can be very environmentally damaging.

- Zinc stocks are expected to be depleted within decades at current rates of exploitation. Extraction releases cadmium and other heavy metals which are very damaging to the environment, and the only practical way of reducing this is to limit consumption. Zinc leaches out into the environment from sheet zinc and galvanized surfaces. The environmental impact of this is still the subject of discussion. Zinc can be recycled but this is not economically viable at present.

- Lead is in extremely limited supply. It is a hazardous material: its production and use in paint and roofing all produce pollution. Almost all sheet-lead is recycled.

- Copper in pipes and on roofs causes pollution and can kill water organisms. Copper is long-lasting; it is economically attractive to recycle it, and this takes place on a large scale.

Synthetics

The basic raw material is petroleum, which is for all practical purposes an irreplaceable resource; Greenpeace estimates that stocks are expected to be depleted within 40 years. It seems likely that consumption is more likely to be limited by the resultant pollution than by depletion itself. Extraction and transport have caused environmental disasters. Refining and manufacture use energy, release organic hydrocarbons and create waste. Harmful substances are often added to create specific properties. Synthetics do not generally cause problems during use, but after demolition. Dumped waste does not degrade, and heavy metals may leach out; incineration generates energy but can also create harmful emissions. Many synthetics can be recycled, but the recycled material must be uncontaminated.

- Polyethylene and polypropylene are simple polymers, the production of which from semi-manufactured products creates little pollution and uses relatively little energy. The additives used are relatively harmless. The materials can be both recycled and incinerated.

- PVC (polyvinyl chloride) requires less energy to produce than many other synthetics, but during its production dioxins, some of the most toxic chemicals known, are created and released. Many additives – such as phthalates linked to cancer, kidney damage and reproductive disorders – are needed, and over their lifetime, PVC products can leak such harmful additives. Furthermore, at the end of their lifetime, PVC products must be either burned or buried. Burning creates and releases more dioxins and other chlorine-containing compounds that contaminate our land and waterways. The toxic gases released when PVC burns have been the principal cause of death and injury in a number of large fires. Dumped PVC does not degrade but does cause leaching of harmful additives such as heavy metals. PVC can be recycled, but in practice it contains such a variety of additives that uncontaminated high-quality material is very hard to produce.

- EPDM (synthetic rubber) causes minimal environmental harm during manufacture, although organic solvents are used to treat the semi-manufactured product, which can cause harm to health and the environment. At the end of its use EPDM can be reused as a membrane if it has been loose-laid and not bonded to the roof deck. Otherwise it can only be ground up into granules for use as filler, which is a low-grade form of recycling.

- One of the principal ingredients of polyurethane is isocyanate, which is extremely hazardous to health. Many of the additives employed are also hazardous. Blowing agents used to make foam products may affect the ozone layer or be hazardous to health. It is difficult to separate from other waste on demolition, and has environmental effects on disposal.

- Polystyrene creates emissions of styrene and benzene during manufacture. Extruded polystyrene requires more energy to produce than expanded polystyrene.

Paints

An important element of most paints is the organic hydrocarbons released, particularly during application, which is estimated to create nearly as much pollution as emissions from vehicle exhausts. They can harm the health of the painters and occupants, and contribute to the level of organic hydrocarbons in the atmosphere which cause smog. Most paints contain harmful additives. Incineration or dumping of painted materials and painting equipment, paint tins etc., releases many of these harmful elements.

- Alkyd paints contain resin and use organic hydrocarbons as solvent.

- High-solids alkyd paints contain less organic solvent.

- Acrylic paints contain less organic solvents than other paints and use water as the principal solvent. They do, however, contain many other harmful substances as anti-corrosion agents and emulsifiers, and there are significant environmental hazards posed by the waste generated during manufacture.

- Natural paints are based on largely vegetable raw materials rather than petroleum. The waste produced during manufacture is generally degradable. They do, however, contain potentially harmful organic hydrocarbons, although they may be from natural sources.

- Mineral paints are also good from an environmental point of view.

- Sealants contain many similar compounds to paints. Many additives are toxic or can cause sensitivity reactions. Many sealants, those based on polyurethane for example, may involve harmful production processes. Sealants based on natural organic products are available and avoid these problems.

- Specifying materials which do not require decorative or protective finishes – self-coloured render or hardwood cladding for example – is generally environmentally preferable.

- Conventional alkyd paint, red lead, epoxy systems and unpainted galvanizing are to be avoided.

Timber

This is the most important renewable raw material used in construction and its use is described in some detail in the next section.

Costs and benefits

The sustainable specification of materials will reduce the depletion of resources and will reduce the environmental impacts from pollution. Some low-environmental-impact materials may cost more, but many cost much less because they use less resources and are used in or near their natural state.

Summary: specifying materials to reduce environmental impacts

The energy 'embodied' in the construction of the dwelling itself is a significant proportion of the total energy used and resulting emissions during the lifetime of a low-energy building. Where possible, specify materials which are:

- Second-hand

- Natural and/or minimally processed

- Locally produced.

Materials should be specified which have minimum environmental impacts. This can be established by considering the Life Cycle Analysis of the material. The Green Guide to Housing Specification is an accessible presentation of this type of information.

Avoid specifying materials (polyurethane and polyisocyanurate insulation in particular) which have a Global Warming Potential greater than 5.

The sustainable use of timber

This section outlines the uses for timber and other wood-based products, and measures to reduce consumption. It describes the potential for UK-grown timber, the properties of imported softwood and hardwood, and certification of timber from sustainable sources. Preferences in the use of timber are outlined.

The use of timber and wood products

As has been said, this is the most important renewable raw material used in construction. Production requires little energy and produces little pollution. The environmental effects arise from poor forestry management, preservative treatment to prevent decay, and the transport distance from forest to building site.

Durable timber

Durable wood does not require treatment and avoids the environmental hazards associated with timber treatments which are discussed below. Tropical hardwoods have evolved to resist insects, fungi and moisture, and are very durable. Many temperate hardwoods are also described as durable. Specifying tropical wood from sustainable sources may have beneficial economic and social effects. However, relatively few sources of sustainable tropical wood exist. The only reliable guarantee is wood certified by the Forestry Stewardship Council (FSC). Timber certification is described below. Temperate forests in North America, Siberia and Australia are often not sustainably managed. European forests, on the other hand, tend to be better managed and are closer to hand. However, relatively little durable wood is produced. Red Cedar from North America and Oak from Europe (including the UK) are durable.

Oregon Pine from North America and Douglas Fir and Larch from Europe (including the UK) are described as 'moderately durable', and will also avoid decay if sapwood is avoided, if the timber elements are carefully detailed to avoid constantly moist conditions, and if maintenance is good.

Most timber used for construction purposes is softwood, which is susceptible to decay if it is constantly damp: this permits fungus and insects to attack. The common response is to treat the timber with toxic chemicals. Alternatives to this approach are outlined below.

Timber treatments

Wood preservatives are generally toxic products that are to be avoided if possible. Workers at treatment plants, site workers handling damp timber or inhaling sawdust from treated timber and the occupants of buildings are all at risk. Treatments variously attack the nervous system, the liver, or are carcinogenic. They may be ingested, absorbed through the skin or inhaled. There has been a tendency for timber treatments to replace the proper specification, design and maintenance of timber in buildings as a means of preventing fungal and insect damage. Chemical treatment is thought by many in the field to be less effective, more expensive and more dangerous than avoiding the problem by good design and specification. To prevent decay and avoid the need for treatment:

- Use durable or moderately durable wood which does not require treatment if the timber elements are carefully detailed to avoid constantly moist conditions and if maintenance is good.

- In new construction, avoid details which cause timber to get wet and not be able to dry out. Timber will not decay if it gets wet but is able to dry out between rain storms (on the outside of a building for example).

- In refurbishment, correct building faults causing timber to get wet and increase ventilation around timber at risk.

- Inspect and maintain buildings regularly.

- It may be necessary to treat wood locally which is at risk, for instance the bottom joints of softwood doors and windows.

- Solid implants of boron provide localized protection. Boron is a fungicide but is virtually harmless to mammals and the environment, unlike other preservatives. The preservative only becomes active if and when the moisture content of the timber reaches 20%, which is when it becomes at risk from decay.

- It should hardly ever be necessary to provide overall timber treatment. Most timber in a properly maintained building is not at risk, and should not be treated. This is now reflected in the revised technical requirements of the National House Builders Council (NHBC), and other building insurance schemes which no longer require interior timber to be treated. There is no reason at all to treat non-structural and decorative timber.

- First consider boron-based treatments if overall treatment is considered necessary. The problem here is that they leach out in water and so are not effective externally unless protected promptly by painting.

- In external situations, CCB (copper, chrome and boron) and ZCF (zinc, copper and fluorine) salts contain fewer harmful elements than other available treatments. New treatment methods are under development, and you may wish to investigate using one of them: non-leachable borates, boron vapour diffusion or alkaline copper salts.

- Water-borne treatments are preferred to organic solvent-based ones, which are very environmentally damaging and hazardous to health; however, water-based treatments cause timber to swell and may lead to warping. For this reason external joinery in particular is routinely treated with organic solvent-based treatments.

- Use the minimum loading of chemicals for the risk situation.

- Specify pre-treatment in the factory rather than site application of preservative – which is the least effective and most hazardous method.

- Avoid treatments that contain CCA (copper, chrome and arsenic) salts. These are the most common and go under various trade names. Also avoid creosote, permethrin, lindane, PCP (Penta Chloro Phenol), TBTO (Tri Butyl Tin Oxide) and bifluoride compounds.

- In refurbishment, ensure that surveys for rot and insect damage are carried out by surveyors who are independent of the treatment companies and who hold the Certificate of Timber Infestation Surveyors awarded by the Institute of Wood Sciences. A *Building Trades Journal* survey reported 'staggering incompetence' in some companies, who specified treatment for non-existent decay while missing genuine infestations.

- You will find that many timber components such as windows are manufactured with treated timber as standard, and it is often difficult to avoid treated timber in these circumstances.

Timber-panel products

Chipboard and MDF (Medium Density Fibreboard) are made from low-grade and waste wood. OSB (Oriented Strand Board) requires higher quality wood chips, and the veneers needed for plywood require good quality roundwood. The bonding agents used generate pollution. Formaldehyde can be released into dwellings, although plywood, chipboard and MDF are available with minimal or no risk of this contamination. Hardboard and Medium Board, which is a structural board similar to hardboard, have low environmental impact as they are manufactured from waste sawdust bonded at high pressure and temperature by the resins occurring in the timber naturally. Softboard or fibreboard are softer versions, but are often impregnated with bitumen, a potentially hazardous material, for use in construction.

Cement- and gypsum-bonded fibreboards and wood-wool slabs have moderately high energy content, particularly from the cement content, and the extraction of raw materials causes environmental damage.

Synthetic resin boards used for cladding take more energy to produce than other board materials and contain a high level of resin for bonding. They are, however, maintenance-free and more resistant to impact damage than cement-bonded fibreboard for example.

Timber consumption

Deforestation is a matter of global concern. An area around the size of the UK is deforested every year, although much of this is as a consequence of clearance for agriculture and cutting trees for firewood. It causes soil erosion and flooding from storm-water runoff. The loss of tropical forests is of particular concern because of the loss of biodiversity and around 1,000 species of tree are threatened with extinction from logging. The consumption of timber in the UK is around 1 tonne per person per annum and rising quite sharply. The construction industry uses nearly three-quarters of the timber used in the UK, and the housing sector is the most significant use within that.

Conservation, waste reduction, reuse and recycling

Consumption can be reduced by using engineered timber, including laminated and composite members, although these may rely on glues which are environmentally damaging. There is also a huge amount of waste involved in the use of timber in Britain, estimated to amount to between 10 and 15 million tonnes per annum. By no means all arises in construction, but nevertheless it is estimated that around 0.75m tonnes of timber could be recovered from demolition sites for reuse. The quality of reclaimed timber is often very good: the material will be well seasoned and tends to have been cut from mature trees with straight, close grain and few knots. However, the market in reused timber is very underdeveloped. There is a limited market in quality architectural salvage, which provides a source of second-hand period features and high-class finishes, but it is expensive. We should aim for a change in attitude in favour of reuse in preference to using new timber.

UK-grown timber

10% of the land area of the UK is forest, compared with an average of around 25% for other European countries. British forests absorb around 2% of the CO_2 produced in Britain, as compared with Sweden where the forests absorb 145% of the nation's CO_2 production. Nevertheless, British forests currently produce 15% of the timber we use, and this is set to rise to around 25% over the next two decades or so.

About one third of home-grown timber production is hardwood used largely in the furniture industry. Oak and Sweet Chestnut, however, are good timbers for construction purposes. Most home-grown softwood is spruce or pine, and is used for relatively low-grade uses such as pallets, fencing, paper and chipboard manufacture. Some is used for non-structural framing, and chemical preservatives are often used to extend its useful life. However, two species grow well in Britain which are good for structural purposes, Douglas Fir and European Larch. They can be structurally graded and kiln-dried to provide good structural material which is classified as moderately durable and less susceptible to decay than most softwoods.

Imported timber

However, the vast majority of timber used in construction in Britain is imported from temperate forests in North America, Scandinavia and Russia They generally have established management, production and replanting policies. The area of such forests is increasing, so that there is now twice as much wood in Europe's forests as there was 100 years ago. Nevertheless, they consist largely of dense conifer plantations which are clear-felled, leading to a very biologically impoverished environment.

A certain amount of tropical hardwood is imported into Britain for use in the construction industry. Common uses are stair strings, door lippings, sub-frames for metal or PVC windows, hardwood doors and windows and plywood – particularly from the Far East, Philippines, Malaysia and Indonesia. It is estimated that over half the tropical timber imported into the UK is cut illegally at source.

Certification

Some consumers have been playing safe by not using tropical timber at all. There is an argument, however, that this discriminates against less-developed national and local economies where timber may be one of the few sources of revenue available. The demand for high-quality timber also puts more pressure on old-growth forests in the temperate regions with the attendant loss of biodiversity. What has been needed is a worldwide independent certification system which can guarantee the sustainable sourcing of timber. The Forest Stewardship Council (FSC) is the most significant scheme. It is an international not-for-profit umbrella group for certification organizations with members who are timber traders, environmentalists, indigenous peoples' organizations, community forestry groups, forest workers' unions, forestry professionals and retail companies. However, FSC-certified timber currently

Look for this logo of the Forestry Stewardship Council.

accounts for around no more than 0.5% of the world's forests. There are a number of FSC-certified forests in Britain. Many are certified under the UK Woodland Assurance Scheme (UKWAS) which is one of a number of national schemes.

The FSC approach has been criticized recently by environmentalists, because they allege that certificates go to large logging companies, whose roads open up areas of primary forest to exploitation in other ways, do not provide stable local employment, and use their green credentials to lever open markets in the US and Europe to illegal imports as well as certified timber.

A number of other certification schemes have been established some of which are recognized by the Central Point of Expertise on Timber (CPET), an organization set up by the UK government to develop and implement their sustainable timber procurement policy, after a number of embarrassing high-profile demonstrations by environmentalists have highlighted the continuing use of illegally logged timber in new public buildings.

The five certification schemes recognized by the UK Government's Central Point of Expertise on Timber are:

- Forestry Stewardship Council (FSC)

- Programme for the Endorsement of Forest Certification (PEFC), which currently recognizes 13 European national schemes under one umbrella

- Canadian Standards Association (CSA) a not-for-profit, independent body

- Malaysian Timber Certification Council (MTCC), also independent and not-for profit

- Sustainable Forestry Initiative (SFI), which covers many forests in North America.

There is controversy regarding the standards imposed by the last three schemes, and some

recommend that timber from these sources should not be used until the timber industries concerned operate in a less exploitative manner. Meanwhile, avoid Government-sponsored or trade schemes.

The Convention on International Trade in Endangered Species (CITES) lists a few species of trees which are commercially significant as endangered.

Cost Implications

There is a marginal additional cost using second-hand or certified timber, but it is difficult to quantify. The sustainable specification of timber conserves natural resources and biodiversity, and the use of home-grown timber promotes the rural economy.

Summary: the sustainable specification of timber

Forests can be conserved and good UK forest management can be encouraged by the following measures in order of preference:

- reuse second-hand timber

- use timber composites and panel products which contain 75% or more recycled material

- use home-grown certified timber

- if you are unable to source home-grown certified material, use uncertified material and satisfy yourself that the forestry management practices used are acceptable

- finally, use certified imported timber

- above all, avoid threatened species.

Avoid using treated timber: there are environmental risks at treatment plants, risks on-site from handling or inhaling sawdust, and health risks to residents from potentially toxic airborne chemicals.

Reuse and recyclability

The construction industry generates a great deal of waste, and it is estimated that up to half the waste going to landfill is from construction. Promoting recycling in the construction industry should be a priority, as the Government is committed to meeting the EU Waste Directive, which requires Britain to substantially increase the proportion of waste which is recycled. This section considers the use of reclaimed and recycled materials in construction and design for future recycling.

The reuse of buildings

All new building is going to be damaging to some extent in strictly environmental terms, and so you should only build new when all possibilities of refurbishing and reusing existing buildings have been exhausted. Prefabricated timber buildings are routinely taken down and reused in North America and elsewhere, and there have been examples in

> *Only build new when all possibilities of refurbishing and reusing existing buildings have been exhausted.*

This Segal Method house was taken down and moved from Glasgow to a new site in Fife in the east of Scotland. The original arrangement included a large conservatory, but the plan was substantially amended when the house was rebuilt.

Britain. One example is the Segal Method of timber construction, based on the use of standard building products used in their standard sizes within a modular framework, and one or two Segal Method buildings have been taken down and the materials reused to build new buildings.

Reuse of materials in new buildings

When designing new buildings:

- Consider the potential for using second-hand materials in the construction

- Give preference to new materials which can be easily recycled and which have a high recycled content

- Design in a way that permits the reuse or recycling of materials at the end of the building's useful life.

Limits on the use of reclaimed materials

There are a number of issues which act as disincentives to the growth in the use of reclaimed materials (materials which can be reused with little or no processing – second-hand bricks, for example):

Left: The owner saw the wood blocks being thrown out, but they make a really good-looking floor.
Right: Second-hand bricks often have much more character than machine-made new bricks. However, second-hand materials may not be readily available in all areas.

- The market for second-hand materials is under-developed in Britain. There is a small market for architectural salvage – high-quality items, but at a high cost. There is an established market in certain materials in certain areas (bricks in and around London, for example, whereas this market does not exist to any great extent in the North of England)

- The development of a market in general materials is restrained by the common perception that second-hand materials are inferior. In practice, the reverse is often the case. Reclaimed timber is generally well-seasoned timber cut from mature trees with close, straight grain and few knots. Second-hand bricks and tiles are often hand-made with a depth of character and texture which a machine-made article can never attain

- The market is limited by the general reliance on standards, quality assurance, guarantees and so on in the construction industry. It is often difficult to give such guarantees on second-hand materials – although as we have said, they may often be of superior quality

- Information on the availability of reclaimed materials is also very limited.

Transporting reclaimed materials long distances increases their environmental impact, yet it is surprising how far you can carry a material for it still to be environmentally better than a new material. The maximum distances you can carry reclaimed materials before they will have a greater environmental impact than new materials produced locally are quoted from the BRE Green Guide to Housing Specification below:

- Reclaimed tiles   100 miles
- Reclaimed aggregates   150 miles
- Reclaimed bricks   250 miles
- Reclaimed slates   300 miles
- Reclaimed timber   1,000 miles

Reclaimed timber is always worthwhile.

Pre-consumer recycled materials

Once you have considered reclaimed materials you should look at recycled materials which require reprocessing before reuse. Many materials routinely contain a proportion of recycled waste generated during production processes (so-called pre-consumer waste); an example is chipboard, made with a high proportion of timber waste from converting trees into sawn or planed lumber. Recycling post-consumer waste, that is waste produced after something has been used, after a building has been demolished for example, generally has a greater environmental benefit.

Post-consumer recycled materials

A number of post-consumer recycled materials are routinely used in construction: granular fill used as a sub-base for roads and pavements, and hardcore is generally crushed brick or concrete. Up to 20% recycled aggregates can also be used in concrete, which will still retain its technical specification. A number of plastic products are made from recycled plastic, which include timber substitute, street furniture and cladding boards. It is also possible to recycle components such as boilers, lift motors and electrical equipment.

The Green Guide to Housing Specification

This gives information on a number of measures relating to recycling:

- Recycled input: the quantity of recycled or waste material contained within a product

- Recyclability: the percentage of the material which can be recycled or reused at the end of the life of the product

- Currently recycled: the proportion of the product currently recycled or reused

- Energy saved by recycling: comparison of the energy required to recycle and/or reuse the product compared with the energy required to produce a similar product from primary resources.

This information is presented as an A, B or C rating, A being the best.

Design for recycling

Design for recycling starts with specifying materials which are easy to reuse or recycle as outlined above. The way that they are combined is critical. A number of common-sense measures apply:

- Avoid composite materials (bonding aluminium onto plywood, for example)

- Reduce the amount of Portland cement used in mortars and renders – or better, use lime-based mortars and renders so that bricks can be reused

- Use mechanical fixings in preference to adhesives; bolt steel rather than weld it

- Adopt dry construction techniques wherever possible (e.g. use neoprene gaskets to seal glazing rather than mastic).

Costs and benefits

Using recycled or reused materials will reduce environmental impacts and waste. Some second-hand materials may cost more but many will be substantially less than new materials.

Summary: using recycled or reused materials

When designing new buildings:

- **Consider the potential for using second-hand materials in the construction.**

- **Give preference to new materials which can be easily recycled and which have a high recycled content.**

- **Design in a way that permits the reuse or recycling of materials at the end of the buildings useful life.**

The Green Guide to Housing Specification **gives useful information on the recycled content and potential for recycling materials.**

Reducing potentially harmful impacts on health

This section gives advice on thermal comfort, providing adequate daylight, sound insulation, and avoiding potentially toxic pollutants and electromagnetic radiation within the home.

Thermal comfort

This section outlines the factors that affect thermal comfort and describes how manipulating these can improve comfort and save energy.

Factors affecting thermal comfort

Comfort is the product of the complex interrelation of a number of factors which take into account among other things what you are doing, whether you are asleep or working hard, and what you are wearing – this might be a T-shirt only, or you may have a number of layers with a thick jumper. Consideration of the balance between these factors can create more comfortable conditions and also create comfort at lower cost and expenditure of energy.

- *Air temperature* This we all know about; it is measured with a thermometer.

- *Mean radiant temperature* This is related to air temperature, and just as critical to comfort. It is the mean of the surface temperatures of surrounding surfaces weighted by area, emissivity (the ability to radiate heat) and proximity. The radiant temperature of a single-glazed window is low on a cold day, which explains why we feel cold next to a large window in a room that is otherwise warm. We can also feel warm sitting in the sun in a room which has a relatively low air temperature. Comfortable conditions will be created with a lower air temperature if the radiant temperature is higher, as is the case with double-glazing and good thermal insulation. This leads to energy savings. Thermal shutters reduce heat loss but also increase the mean radiant temperature.

- *Air movement* An increase in air speed will lead to an increase in the rate at which the body loses heat by convection and evaporation. High air movement is essential to keep cool in hot conditions without air conditioning, but will cause a feeling of uncomfortable cold draughts in a heated room. Reducing unwanted air movement by draughtstripping is necessary in winter, whereas designing for good cross-ventilation and promoting the natural stack effect of warm air rising and being replaced with cool air at low level are important in summer.

- *Humidity* High humidity can be a problem in hot climates, where it reduces the ability of the body to shed heat through evaporation. Low humidity can be a problem in a cold climate in winter, when external very dry air comes into a warm house and causes the relative humidity to drop. It can also be an issue in air-conditioned buildings, where low relative humidity makes our body lose more moisture through evaporation, leading to the symptoms of dehydration: headaches, dry throat and irritated eyes. If the relative humidity is above around 70%, it can lead to mould growth and respiratory problems; if it is below 40%, it can lead to dehydration. Humidity can be modified by ventilation, most importantly, but also by using hygroscopic materials on the inside of buildings. These are materials which absorb moisture into the pores of the material and store it as liquid water when the humidity is high and releasing the moisture again when the relative humidity is low, thus evening out swings in the relative humidity inside the building. Materials which behave in this manner include lime plaster and lime mortar, timber, cork and fibreboards. Vapour-permeable finishes should be used. Moderately hygroscopic materials include plywood, chipboard and gypsum plaster. Materials with no hygroscopic qualities include glass, plastic, glazed tiles, concrete and metals. A vapour-balanced construction or 'breathing construction' will tend to reduce high levels of humidity as vapour migrates through the construction to the outside of the building.

> *Providing a house is well insulated and draught-proofed with double-glazed, openable windows, most people will be comfortable most of the time.*

The balance between these factors is more critical in situations where the occupants do not have much control over the situation: in an air-conditioned office for example. In a house, however, there is a relatively high level of control over temperature and ventilation. Providing a house is well insulated and draught-proofed with double-glazed, openable windows, most people will be comfortable most of the time.

It has also been shown that a passive solar design where temperatures may fluctuate somewhat generates a comfort tolerance into its occupants. Providing temperature changes are gradual, we will dress more appropriately for the season and not expect to be able to wear no more than a T-shirt all year round.

Summary: improving comfort and reducing energy use

Consider the balance of factors that affect thermal comfort to reduce heating costs and improve comfort.

Daylight

This section considers sizing windows to balance the need for daylight with the need to reduce heat loss, the design of rooms for good lighting, and how to calculate for adequate daylight.

Windows, daylight and energy

Plentiful daylight is important to a sense of well-being – so too is sunlight. Good daylight can also save energy. However, a balance must be struck between saving from electric lighting and heat loss through windows, which will be up to six times as much as the heat loss through a solid wall. The potential energy-saving is greatest in buildings occupied largely during the day, such as offices and schools. Daylight is less important in a house

- *Activity* The rate at which our body converts food into heat (our metabolic rate) is about four times higher when we are doing heavy work than when we are asleep. This is why we need lower temperatures when we are active than when we are sitting still or asleep.

- *Clothing* We can modify the balance between our body and our surroundings by adjusting the clothes we wear, which will affect how warm or cold we feel. The range extends from a warm winter coat or a duvet to no insulation when we are naked under the shower. To be comfortable, the temperature needs to be warmer in the bathroom (around $27^{\circ}C$) than in the bedroom (around $18^{\circ}C$). The temperature in the bathroom is also more critical – no more than $1^{\circ}C$ either way, as compared with the bedroom which can be up to $3^{\circ}C$ warmer or colder without causing discomfort.

occupied and heated largely in the evening, and even less important for the bedrooms. The energy-saving can be maximized if insulating shutters or curtains are used over windows at night.

So what is the appropriate size for windows? It's probably more relevant to ask the question as: How small can we make the windows while still achieving adequate daylight? A number of simple observations will help. Rooflights are more than twice as effective as windows in admitting light, because they face almost unobstructed sky, whereas a window tends only to 'see' sky over about one third of its area. Light-coloured interior surfaces will reflect light more effectively within a room.

Light from two sides

A room which has more than one window will have a better pattern of light distribution, in contrast to a room which is deep and narrow with a window in the short side. This is a common situation which leads to an uncomfortable contrast or glare near the window and gloom away from it. A rule of thumb is that a room can be no more than 6m deep to be effectively daylit. Observation shows that rooms lit from two directions create less glare around people and objects, and allow us to perceive the form of surfaces properly. Such rooms will also enjoy sunlight at different times of the day. It is relatively easy to design for this in a small building, but in a large building or terrace of

houses some articulation of the plan will be necessary. Even so, it is surprising how many windowless flank walls are to be seen. If light from two sides cannot be achieved, plan shallow rooms with two windows or more in the long wall, or at worst, if a deep room is necessary, make the ceiling high with a large window with deep, splayed reveals to reduce glare.

Calculating daylight

The calculation of Daylight Factors places this on a scientific basis. The Daylight Factor is the amount of daylight falling on a horizontal surface in a room as a percentage of the daylight falling on a horizontal surface outside.

Average daylight factor calculation

Average daylight factor in rooms with a side window or rooflights:

Daylight factor % = MWßT / A(1 − R²), where
W = total glazed area of windows or rooflights
A = total area of all the room surfaces (ceiling, floor, walls and windows)
R = area-weighted average reflectance of the room surfaces
M = a correction factor for dirt
T = glass transmission factor
ß = angle of visible sky (see diagram)

Guide values for a typical dwelling with light-coloured walls are as follows:

R = 0.5
M = 1.0 (vertical glazing that can be cleaned easily), 0.8 (sloping glazing), 0.7 (horizontal glazing)
T = 0.7 (double-glazing), 0.6 (double-glazing with low-emissivity coating)
ß = 65⁰ (vertical glazing)

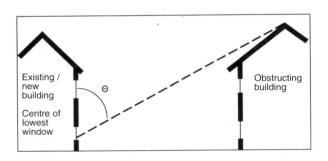

Above: Diagram showing the angle of visible sky.

But be aware that most people will find the minimum levels suggested in the British Standard and required under EcoHomes too gloomy. Most people will feel that levels about twice the minimum are better. The minimum recommendations are:

Kitchen	2%
Living rooms	1.5%
Bedrooms	1%

The Building Regulations used to require that a habitable room had a window the size which is at least 10% of the floor area. This gives a daylight factor of between 1 and 1.5 for an average domestic-size room of 12 square metres in area. A daylight factor of 2 requires a window of about 15% of the floor area and a daylight factor of 3 requires a window around 25% of the floor area.

The British Standard on daylight includes two other standards in addition to the daylight factor requirement:

- that there is a view of the sky from tabletop height (0.85m) in at least 80% of the area of the room, and

- that the sky can be seen from every fixed work surface and table in the kitchen

Costs and benefits

Windows cost more than walls, but contribute to long-term savings in lighting costs. The proper design of windows can improve the visual environment whilst reducing energy use.

Summary: window design for daylight

Design windows to provide adequate daylight and save energy on lighting whilst achieving a balance with heat loss.

Noise

Few environmental issues adversely affect health more than excessive noise nuisance. This summary deals with noise within and between homes.

House planning

Plan to avoid habitable rooms in dwellings adjacent to communal areas such as shared stair access, avoid bathrooms or living areas adjacent to neighbours' bedrooms, and make sure that noisy plant or lift machinery is more than 3m from a window.

It is also important to consider the vertical relationships between rooms when planning flats. It is best to stack similar rooms above one another – bedrooms above bedrooms, and so on.

When planning a house, it is useful to consider different zones for the family, the parents and the children, each with their own requirements as regards noise, and to plan them in different parts of the house.

Detail design and standards of construction

Sound insulation is a complex subject which depends on careful detailed design and a high standard of construction for success. After you have considered planning to avoid noise nuisance, you will have to consider the sound reduction required from the elements of construction of the dwelling to provide acceptable conditions. There are three issues which you have to consider to achieve sound reduction:

- Providing heavy, massive construction such as brick or block masonry

- Isolating elements of construction so that the structure does not form a path for sound to travel along

When planning a house, consider different zones for the family, the parents and the children, each with their own requirements as regards noise.

Summary: avoiding noise problems

Plan to avoid noise problems from communal areas and between dwellings, and detail the construction to ensure either structural separation or massive construction; avoid air paths through the construction.

Toxicity

This section outlines some of the potential threats to health which can exist in modern construction, and suggests that designers adopt a precautionary position and avoid exposure as far as possible.

The precautionary position

In developed countries we spend 95% of our time in buildings, and we should expect them to be healthy and enhance our well-being. However, we are subjected to gases and particles given off by building materials and appliances within the home, and the effect of this cocktail of compounds on different individuals varies. Most of us adapt, but some people's immune system becomes overloaded leading to increasingly allergic reactions. Other compounds cause birth defects, cancer, disease or poisoning.

The safe policy is to adopt a precautionary position and avoid exposure to those substances which have become part of our environment to which we are not adapted.

Pollutants within the home

- *Nitrous oxide and carbon monoxide* from faulty combustion of gas and paraffin appliances. Fit a gas alarm

- *Radon*, a radioactive gas given off from granite rock in the ground in certain parts of Britain. Provide a gas barrier in the ground floor and

- Making sure that there are no cracks or other air paths for sound to pass through.

In masonry construction, make sure that the density of blocks is sufficient; on site, make sure that joints in walls are fully filled with mortar and cavities not bridged. In timber construction, ensure that there are no structural connections across double party walls or back-to-back electrical accessories forming a weak point in wall constructions.

Sound tests can help to identify problems.

Costs and benefits

Good sound insulation improves comfort and avoids nuisance, and is largely a question of careful detail design and careful execution on site. There may therefore be little or no extra cost involved.

ventilate below. This is regulated by the Building Regulations.

- *Volatile Organic Compounds* (VOCs). A wide class of compounds that includes organochlorines, which are a broad range of plastic compounds including polyvinylchloride (PVC) and polychlorinated biphenyls (PCBs), which are found in many building and household products such as cleaning products and paints.

- *PVC* is relatively inert in use (but see How to Specify Green, Chapter 15) but will off-gas its constituent chemicals for some time after manufacture. These include vinyl chloride monomer (VCM), which is extremely toxic and is carcinogenic. Phthalates are widely used as plasticizers. They are almost universal environmental contaminants and are oestrogen mimics or feminizing agents, and it has been suggested that they may cause falling sperm counts and rising male infertility. They can also leach out into drinking water. The real hazard is if there is a fire, in which case dioxins, hydrogen chloride (a very

In developed countries we spend 95% of our time in buildings, and we should expect them to be healthy and enhance our well-being

corrosive gas) and heavy metals are given off, which together cause acid burns on inhalation, poisoning, cancer, immune system damage and hormone disruption. There are alternatives to all applications of PVC in buildings: galvanized steel gutters and downpipes, clay for drains, polyethylene for above-ground waste pipework, linoleum for flooring, timber for windows and low-smoke electrical cable instead of PVC insulated cable.

- *Paints based on solvents* contain a myriad ingredients, many of which are potentially harmful. Prolonged exposure to hydrocarbon solvents can lead to tremors, loss of co-ordination and depression, and professional painters suffer from high rates of lung cancer. Even when paint is dry it will continue to off-gas VOCs. Most shades of synthetic paint contain titanium dioxide white pigment, which is a possible carcinogen and can cause respiratory problems and skin irritation. Use organic paints, stains and varnishes based on natural oils and resins and earth pigments.

- *Formaldehyde* is present in many building products as a constituent of glue, and the vapour is given off from many synthetic products. It is also present in small quantities in new wood. Formaldehyde is very irritating to skin, eyes and the respiratory system, and may cause cancer. Use alternative glues and zero-formaldehyde versions of chipboard and MDF.

- *Wood preservatives* are one of the most significant sources of toxic compounds in the home. How to avoid their use in described in the previous chapter under Timber Specification.

- *Fungal spores, bacteria and dust mites* can all trigger allergic reactions and can be eliminated by proper ventilation and insulation. Floorboards and linoleum are better than carpets.

Costs and benefits

Many airborne pollutants can be avoided by careful specification of materials at little or no extra cost. Avoiding timber treatments may reduce costs, but many natural finishes and linoleum are more costly.

Summary: avoiding risks to health

Most potential threats to health can and should be avoided:

- Provide a radon barrier in the ground floor construction where there is a risk – the principal areas of risk are in Cornwall and Devon.

- Specify natural paints, stains, mastics and adhesives to avoid VOCs.

- Specify galvanized steel gutters and downpipes, clay for drains, polyethylene for above-ground waste pipework, linoleum for flooring, timber for windows and low-smoke electrical cable to avoid PVC.

- Specify formaldehyde-free board products.

- Avoid timber treatments by proper design and specification.

- Provide good ventilation to prevent mould and dust mites causing allergic reactions.

Pollution

This section outlines how to reduce pollution emitted from homes from domestic boilers.

Low NOx boilers

Low pollution boilers are designed to reduce the emissions of oxides of nitrogen NOx, and they are also referred to as low NOx boilers. Nitric Oxide NO and Nitrogen Dioxide NO_2 are produced from the incomplete combustion of fuels, and gas in particular. They are greenhouse gases and therefore contribute to global warming. They also react in the atmosphere with sunlight to produce ozone. Whilst ozone is necessary at very high level in the stratosphere to protect the earth from radiation, at low level it is a pollutant which leads to increased asthma and other respiratory complaints, and to damage to crops. Nitric oxide oxidizes in the atmosphere to form nitrogen dioxide which in turn forms nitric acid which is the principal ingredient of acid rain.

Some boiler manufacturers have designed their products to reduce NOx emissions by designing the burner to pre-mix the gas and air before combustion and by careful design of the combustion chamber. British Standard BS EN 483 for gas-fired central heating boilers sets five classes for maximum NOx emissions. In fact most gas boilers achieve the highest standard, although some boilers have a much better performance than others, down to 20 mg/kWh.

Oxides of nitrogen are emitted from power stations, so heating by electricity supplied from the grid does imply emissions of NOx; however electricity from wind and photovoltaic panels does not. Electricity and heating from a Combined Heat and Power (CHP) plant or heat from burning wood may be acceptable, depending on the performance of the appliance.

Reducing emissions by using a low NOx boiler

The level of emissions is not related to the cost of a boiler, but make sure that you use a low NOx boiler with emissions no more than permissible under class 5 of the British Standard and at the low end of that range if possible.

Electromagnetic fields

This section outlines potential hazards from exposure to electromagnetic fields within the home and suggests how to avoid them.

Electric fields and magnetic fields

Whenever electricity is used an electromagnetic field (EMF) is produced. There are two types of field produced: electric fields resulting from a voltage differential, and magnetic fields arising from a flow of current. Both diminish rapidly with increasing distance from the source. Electric fields can be screened by most common building materials, whereas magnetic fields will pass through most materials. There is concern, but no conclusive evidence either way, that continued exposure to magnetic fields is harmful and in particular carcinogenic. There is no safe level set in Britain.

Magnetic fields and how to avoid them

The electrical system in a house can cause magnetic fields in two ways:

- *From defective ring circuits.* If one cable becomes unconnected at a socket, for instance, power will be supplied 'the other way round' the ring circuit and so you would never be aware of the discontinuity. However, the imbalance in the circuit will produce an EMF. This can be prevented by using spur circuits in preference to ring circuits, especially serving bedrooms. Sheathed cable which prevents magnetic fields emanating from the cable is available, but it is expensive. Another approach is to fit all circuits with a demand switch which cuts off the circuit when all appliances are switched off for the night – but remember that an increasing number of appliances require power to run clocks etc.

- The other possible source of magnetic fields is the meter and consumer unit. This should not be sited where people are going to sleep.

Some appliances are also a cause for concern:

- A clock radio next to the bed: move it across the room

- An electric blanket: use it to warm the bed, but don't sleep under it

- TV or computer monitor: keep one metre away, and use a radiation screen.

Sources of EMFs outside the dwelling are considered in Chapter 5: Environmental issues and the site.

Costs and benefits

The risk can be avoided at little extra cost for spur rather than ring circuits, although sheathed cable is substantially more expensive.

Reduce the potential risk to health from electromagnetic fields

Reduce the potential risk to health from electromagnetic fields by avoiding ring circuits or using sheathed cable for the electrical installation.

Reducing waste

This section outlines measures that can be undertaken to reduce the amount of waste generated during construction, and how to encourage the recycling of domestic waste.

Construction waste

This section outlines the measures that can be taken to reduce the very significant impact of the waste generated by construction.

The scale of construction waste

The construction industry generates more controlled waste than any other sector in Britain: one third of solid waste, amounting to 78m tonnes. Of this, one third – crushed concrete and hardcore from demolition and road scalpings and the like – is recycled for use as aggregate and fill. It is estimated that only 4% is reclaimed or recycled for high-grade uses.

Meanwhile, the extraction of 250-300m tonnes of material for aggregate for concrete, crushed rock for road-making and clay for brick-making has substantial environmental impacts.

The construction industry generates more controlled waste than any other sector in Britain: one-third of solid waste, amounting to 78m tonnes.

Government targets

Minimizing construction waste and increasing the use of recycled material are both government priorities under pressure from the EU Landfill Directive and elsewhere. An Aggregates Levy was introduced in 2002 which taxes primary aggregates, with the aim of encouraging the use of recycled material. 17% of the aggregates used in 1999 were recycled.

Waste can be dealt with by, in descending order of preference:

- Reusing buildings rather than demolishing them. This also reduces resource inputs into new construction.

- Implementing waste-minimization measures.

- The reuse of materials

- Recycling materials

- Burning waste for energy

- Dumping the residue in landfill.

Waste minimization

Either you, or your contractor if you are employing one, should prepare a Waste Strategy. This would include a waste audit to identify the waste streams and make proposals for dealing with them. This may include waste segregation on site with separate bins for timber, plastics and metals. The BRE have developed SMARTwaste as a web-based tool to monitor waste as it is generated on site so that measures can be taken straight away to reduce it.

- Review the design of buildings from the point of view of buildability and waste at an early stage.

- If you are having a building demolished to clear your site, consider what materials can be reused and recycled. Also consider what materials for

Construction creates one-third of all solid waste in Britain – more than any other sector of the economy.

your new house could be second-hand or recycled. Match the two lists as far as possible. Often material can be used for fill, hardcore or aggregate for concrete. You will need to check the standards for the suitability of recycled aggregates. It is also highly desirable if contaminated soil can be recycled on site in less critical areas or as fill under clean material rather than being exported as hazardous waste.

- The use of prefabricated assemblies can reduce waste and the waste that is generated can be reused and recycled more easily under factory conditions.

- The reuse of materials and components is preferred to recycling. Bricks are commonly reused and there is a small market in architectural salvage. It may be difficult to obtain guarantees of performance although many reused materials may be of superior quality to new, timber for instance. The distance reused materials have to be transported may affect viability. It also takes time to source reused materials and information sources are not well developed. Design and construction programmes have to recognize this. See section on Buildings: reuse and recyclability (p187) for further information.

Costs and benefits

Minimizing waste and promoting a more efficient building process should reduce costs, although reused materials may in some cases have a higher initial cost than new. Less waste will reduce both environmental risks and the use of resources.

Reducing construction waste

Waste should be dealt with by, in descending order of preference:

- Reusing buildings rather than demolishing them.
- Preparing a waste strategy
- Reviewing the design from the point of view of buildability and waste
- Using prefabricated assemblies where possible
- Reusing materials
- Recycling materials.

Domestic waste

This section outlines the measures that can be put in place to encourage recycling and composting and thereby reduce the amount of domestic waste that is sent to landfill.

Reducing landfill

In 2003, 29m tonnes of domestic waste were generated, which is slightly less than the previous year. Every household creates an average of around 23 kg of waste per week. In 1995, 85% of that went to landfill, which had dropped to 72% by 2003. 19% of household waste was recycled or composted, and 9% incinerated with recovery of the energy produced. The EU Landfill Directive, which came into force in 2002, requires Britain to increase the proportion of resources recovered from domestic waste from 14% to 67% by 2015. The government target is to recycle or compost 30% by 2010. These are very onerous targets which will require, amongst other things, every incentive for householders to segregate waste and compost organic waste.

Composting

Organic waste comprises around half domestic waste, and dwellings should have a compost bin if they have a garden. Communal composting facilities are provided in some areas by voluntary sector organizations or local authorities. In other areas, local composting facilities and schemes should be considered for dwellings without gardens. Composting requires a volume of around 2m³ per household.

Recycling facilities

Three recycling bins would generally provide capacity to segregate compost, glass and paper. Bins need to be in a convenient position for people to use them. Under-sink containers that fit in the sink unit are very convenient. Other types of waste that are commonly separately collected at local recycling centres are tins, rags and aluminium cans. Some authorities, however, collect all recyclable material from a single box and segregate it back at the depot. The important issue is that there is sufficient space in the home to store material for recycling. A minimum standard of 0.8m^3 storage space is recommended by the 2006 Code for Sustainable Homes.

Costs and benefits

Organic compost returned to the soil will increase fertility and improve soil structure. A reduction in the proportion of domestic waste that goes to landfill will reduce environmental risks and the use of resources. There is a minimal cost for bins and storage space.

Reducing domestic waste

Contact the local authority to establish what local facilities are available for recycling and composting and provide space for bins for storing segregated waste in a convenient place – including under the sink.

Under-sink, multi-compartment recycling bins are convenient and make recycling easy.

Reducing
water consumption

This section considers water conservation and a sustainable approach to storm water and sewerage.

Water conservation

This section describes the need for water conservation and outlines how to achieve this through the design of buildings, the specification of appliances and good management. It also outlines water recycling, rainwater collection and obtaining water from your own borehole.

Water conservation

The demand for water for buildings, industry and agriculture doubled over the 20 years from 1970 to 1990, resulting in water shortages in some parts of Britain over the last few years. The effect of climate change is uncertain, but parts of the country are likely to get less rainfall. Long-term trends suggest that the UK will have wetter winters and drier summers.

Domestic consumption accounts for around 65% of the total. It has been demonstrated that water-efficiency measures can reduce domestic consumption by up to 50%, so water conservation can have a significant effect.

The cost of water has risen steeply over the last decade or so, and is now likely to be the principal running cost of a low-energy house. Water-efficiency measures also reduce energy use and can lead to other benefits.

The first approach should be to use water more efficiently, before considering wastewater recycling or rainwater harvesting.

You can reduce consumption by a combination of:

- designing buildings for conservation
- using water-efficient appliances
- better management and maintenance.

WCs

The largest use of water in older homes is for flushing the WC – over one third of the total. Dual-flush and low-flush WCs can reduce domestic water consumption by up to 20% overall. The Water Supply (Water Fittings) Regulations 1999 require that the maximum flush volume for a new WC is 6 litres. A Scandinavian WC would use 2 litres for a half flush and 4 litres for a full flush.

Quoted flush volumes are nominal and ignore the water that enters the cistern while the WC is flushing. Depending on speed of fill, actual flush volumes can easily be 10-20% higher than the nominal volume. Delayed-action inlet valves are available; they solve this problem without affecting flush performance, and can be retrofitted to most WCs.

The recent 1999 regulations now permit flush valves instead of the previously mandatory siphon. While valves offer several advantages and allow the use of buttons rather than a lever, they will eventually leak and are less robust than the traditional UK siphons with which UK plumbers are more familiar.

Showers

Bathing accounts for around 20% of the water used by a typical UK household, but this is increasing. Generally, showers use around one third as much water as a bath, except for a 'power shower', which can use more than a bath. Also, people tend to shower more often than they bath, which can reduce the saving. 'Water saver' showerheads generate finer droplets and can give the feel of a good shower at around half the flow rate of a conventional shower, but are not to everyone's taste and may create a 'cold feet effect'. They require mains pressure, and can be used with a well-modulated combination boiler but not with a plumbing system gravity-fed from a storage tank in the loft. Thermostatic shower mixers tend to reduce the waste of water that happens while balancing the temperature with separate hot and cold controls.

A shower-flow regulator can be fitted between the mixer and the shower hose to limit the maximum flow rate for power and mains-pressure fed showers: 9 litres per minute should be considered a maximum, and many users will be happy with 6 litres per minute. But beware: instantaneous electric showers should not be restricted, for safety reasons.

Taps

About 8% of domestic water use is at the washbasin. Some 80% of this can be saved by fitting spray taps. This is fine for rinsing hands or toothbrushes, but time-consuming when filling the basin to wash. A new invention, the 'Tapmagic' insert, seeks to overcome this. It can be fitted to most taps with a metric thread or round outlet and at low volume provides a spray pattern. As flow is increased, the device opens to provide unrestricted flow, although this can be splashy. Single-lever ceramic cartridge taps are available with a water-saving cartridge that requires slightly increased pressure force on the lever of the tap to open it for full flow. The best models deliver only cold water in the central position, providing additional energy savings. Taps should be clearly and indelibly marked to identify hot and cold, as otherwise water can be wasted while both are tried.

All new taps should be specified with standard metric outlet threads to allow the option of fitting of aerators, sprays or future innovations.

Flow regulation

Water savings from flow regulation are very variable: regulating the flow to a sink or basin will save water if the users are not conservation-conscious, but will make little difference if they are already in the habit of controlling the flow manually. Regulating the flow to

a high-volume shower will save water, but regulating it to a bath will not. A flow regulator is a device fitted in the pipe serving a tap or shower, but you can control the flow better with spray taps and low-water-use shower fittings (see above). Flow regulators do bring other benefits, however, including more stable pressure throughout the system, in turn leading to more stable temperature and flow for a shower. Flow regulators will provide a relatively constant flow irrespective of pressure, whereas flow restrictors, which are a similar but slightly different type of device, deliver a higher flow if the pressure is higher. A pressure regulator, on the other hand, will give constant pressure irrespective of flow, and is usually fitted at the point where the supply enters the house to limit pressure in a mains-fed hot-water system.

Sprays and aerators are available with built-in flow regulators for ease of installation. Shower regulators are available in a chrome housing that simply screws into the mixer valve before the shower hose and can be fitted without special tools.

Plumbing layout

Low-water-use fittings should have supply pipes no larger than necessary for the flow rate required. Fittings should be as near to the hot-water source as possible to reduce 'dead legs' and the consequent waste from running the hot tap until it gives hot water. Combination boilers can increase the effective dead leg because they take time to warm up when hot water is demanded. Many boilers include a small heat store or a 'keep-warm' device to avoid this problem, so if you intend to install a combination boiler, check the boiler specification.

Appliances

Washing machines use the most water in a house after the WC – around 20% of average consumption. Washing machines and dishwashers now have an 'energy label' which also shows water consumption. The most energy-efficient machines are often the most water-efficient, with typically under half the water consumption of less efficient machines. Energy labelling has driven manufacturers to produce more efficient machines without a price premium, although quality, durability and features will vary. See Chapter 7, Reducing energy in use.

Management issues

While water-efficient systems are crucial, conservation depends largely on how they are used. Fitting water meters to existing houses is often but not always beneficial. It tends to reduce consumption by around 10%. Paying according to metered consumption can reduce the cost of water, particularly for a small household living in a house with a high rateable value.

Make regular checks for leaks and carry out prompt repairs of dripping taps and valves.

Recycling water

Once you have implemented measures to reduce water consumption, you can consider ways of reducing the need for mains water. Waste-water from the bath, shower and washbasin can be collected in an external underground tank, disinfected, pumped to a header tank, and used for flushing the WC. In a domestic situation the amount of waste-water produced is similar to the amount required for toilet flushing, and savings of about 30% are claimed by manufacturers. A trial found that a reduction of around 20% is more realistic, and with efficient WCs this could be even less. The initial cost can be high, leading to long or even negative payback periods when maintenance is included. Existing systems are not entirely robust in their operation and do need maintenance. Current greywater systems don't score well from a lifecycle perspective because they consume chemicals to disinfect the recycled water and use energy to run the necessary pumps. For homes connected to a sealed cesspool, the economics will be better. Larger developments have used on-site treatment of sewage effluent to a quality suitable for reuse in WCs or for landscape irrigation. Such schemes are only economically or ecologically sensible where local water supplies are very limited or large quantities of non-potable water are required, such as for golf-course irrigation.

Rainwater collection

This can be undertaken at three different levels of cost, complexity and saving:

- A water butt is very simple and inexpensive and will offset the use of mains water in the garden, which averages around 5% of total consumption.

- Collecting rainwater for flushing the toilet and for use in the washing machine requires a storage tank of around 2m³ which collects rainwater from the roof, an overflow to the storm drains, a pump and a header tank or other system for providing mains water back-up. This all adds up to moderate expense, but manufacturers claim it can offset up to half of your water use, given sufficient roof area and annual rain. Payback periods for domestic systems are typically over 10 to 20 years, and are made worse by improvements in washing-machine and WC efficiency. Correctly collected and stored, rainwater is generally considered to be safe for WCs and washing machines. If you live in one of the wetter areas of the country, where rainfall is over 2,000mm a year, you should be able to get optimum benefit from such a system with a roof area of around 50m² (about the size of the roof of a small 3-bedroom terraced house). However, rainfall in large parts of the country is around a third of this maximum. In this case the optimum size of collecting area is three times the size of a small house. Rainwater from most roofing materials is 'soft' without dissolved minerals and extends the life of washing machines.

- Collecting rainwater for drinking and cooking requires filtering and purification, and is not generally appropriate, sustainable or cost-effective.

Gardens

Whilst garden use might only average around 5% of domestic water use, this can rise to over 50% in droughts when water is in short supply. Also the potential for wastage due to leaking ponds and water features or poorly managed irrigation systems is considerable. Specialist books provide design guidance and plant lists for low-water gardening.

It is perfectly possible to have a beautiful and productive garden without using any mains water. Use low-water-demand plants and mulch. The lawn is the thirstiest part of a garden – but it will recover from drought conditions on its own without irrigation. A water-efficient garden will be drought-resistant and require less maintenance than a thirsty one.

Boreholes

You may be able to be independent of the mains by having your own supply. A borehole may be cost-effective if it serves a development of several homes. You may also be able to sell water to other neighbouring developments. The supply will have to be tested and approved by the water authority. However, if a mains supply is available it is probably better to use it, as it will probably be the most sustainable option in terms of energy, maintenance, monitoring, initial cost and embodied energy.

Costs and benefits

Many straightforward measures to conserve water have no or low initial cost, and you should experience improved performance: taps that run hot quickly, taps that don't splash, WCs that flush first time, showers with stable flow and temperature, drought-resistant low-maintenance gardens . . . and lower water costs. Rainwater collection and waste-water recycling systems are probably not cost-effective under current conditions, and they both require maintenance. One should probably focus efforts on further energy-efficiency measures rather than on expensive water recycling. Ultimately, however, water recycling will reduce the consumption of a resource which is in increasingly short supply in some areas, notably the south-east of England.

Reducing water consumption

Substantial reductions in the use of water, up to 50%, can be achieved at relatively low cost by:

- Installing low-water-use efficient WCs
- Installing showers with flow regulators
- Providing spray taps or aerators and flow regulators
- Ensuring that the plumbing layout is efficient, with short 'dead legs' to hot-water taps
- Practising good water-conservation management

Once water consumption has been reduced as far as possible, one may consider replacing the use of mains water with either recycled water for WC flushing or rainwater for uses other than cooking and drinking.

Storm water

This section deals with the disposal of rainwater from roofs, roads and other hard surfaces and outlines the Sustainable Urban Drainage Systems (SuDS) approach and describes some of the components of such a system. It also describes some of the implications for the design and approval process for a SuDS scheme.

Problems with storm-water drainage

There has been widespread flooding in the UK in recent years, and the number of water-pollution incidents has risen steeply. The effect of climate change on rainfall is unclear, but the general trend is towards more extreme weather patterns and some parts of Britain are likely to experience an increase in rainfall.

General urban practice is to collect storm water from roofs and roads in drains and dispose of it into a watercourse or a soakaway, often some distance from the site. Concentrating storm water in drains in this way can cause flooding or pollution, and disrupts the natural water cycle. There is a substantial cost in either providing separate storm-water drains and sewers or, as in some areas, notably London, sizing the foul-drainage system to take storm water in a combined system. Usually the sewage-treatment plant at the end of such a combined system can only cope with limited volumes of storm water, so that partly treated sewage is released periodically into the environment.

Planning Policy Guidance PPG 25 aims to ensure that all new development should, as far as possible, incorporate sustainable drainage measures to avoid adding to flood risks elsewhere.

Sustainable Urban Drainage Systems

A Sustainable (Urban) Drainage System (SuDS) is a way of dealing with these issues sustainably. Currently, a conventional piped system is designed to take away surface water to prevent flooding locally. However, it does not deal with the problem of pollutants from roads and elsewhere being washed into rivers or the groundwater where they are difficult to remove. The wider issues of conserving water resources, protecting wildlife habitats and the landscape value of lakes and rivers have also been largely ignored until now.

SuDS aims to treat quality, quantity and amenity issues equally.

SuDS is a more sustainable approach because it:

* protects and can enhance water quality

* can reduce flooding impact

* is sympathetic to the environmental setting

* can provide a wildlife habitat in urban watercourses

* can encourage the replenishment of natural groundwater supplies.

It achieves this because it:

* deals with runoff close to its source

* manages potential flooding at its source

* protects water resources from pollution.

With SuDS, the principal methods of surface-water control slow the rate of flow to prevent flooding and erosion, thereby spreading peak flows over a longer period. They also treat water carrying pollution and dispose of surface water into the soil or into rivers. These elements should be designed as a complete system.

* *Filter strips and swales* These are slopes or shallow channels respectively, with vegetation (often grass), which carry water evenly off an impermeable area. Rainwater runs in sheets through the vegetation, which slows and filters the flow. Swales can carry water away or have check dams to slow the flow and increase infiltration into the soil. The vegetation combats pollution by trapping organic and mineral particles, which are incorporated into the soil and the vegetation encourages biological treatment. The method is particularly suitable for draining residential roads and parking areas.

- *Filter drains and permeable surfaces* These have permeable material such as crushed stone below ground. This stores water which then infiltrates into the soil. A filter drain is a trench filled with permeable material with a perforated pipe at the bottom, and is used for draining water off the side of a road or car park. Permeable surfaces include grass which may be reinforced with a grid of concrete or plastic to take traffic, gravel, block paving designed with gaps, or surfaces such as porous asphalt. Permeable surfaces do not require a complex pattern of falls or gullies and so are quick to construct.

- *Soakaways and infiltration trenches* These enhance the ability of the ground to absorb water. Essentially they are holes or trenches filled with crushed rock or similar material.

- *Basins and ponds* Basins are dry except during wet weather, unlike ponds which contain water all the time. They tend to be at the end of the surface water system, and are used if the water cannot be dispersed at source or for landscape purposes. They can retain floodwater and treat pollution by the settlement of solids, absorption by aquatic vegetation and biological activity.

Green roofs

Green roofs with soil and vegetation slow down runoff. They also soften the visual impact of development and can prevent the building overheating in summer. They can be expensive if they are designed with complex systems to retain moisture in summer. Although an economic, simple roof of vegetation chosen to survive in hot dry conditions – sedum, for instance – will tend to go brown in summer, it will go green again as soon as there is a reasonable quantity of rainfall. The extra weight may increase the cost of the supporting roof structure.

Approvals

Responsibility for drainage is split between the local authority planning and highways departments, building control, the water authorities and the Environment Agency, who give permission to discharge into watercourses. Close consultation between all these parties is essential for the design of an integrated Sustainable Urban Drainage System.

Design team

The design of an integrated drainage system demands a wider range of skills than normal, including knowledge of hydrological flows, landscaping and ecology as well as the engineering of earthworks, wetlands and other features.

> *Green roofs with soil and vegetation slow down runoff. They also soften the visual impact of development and can prevent the building overheating in summer.*

Costs and benefits

Controlling surface water as near its source as possible and without the use of an underground drainage system should reduce initial costs. A modest rebate of around 15-20% is available from some water companies for households that do not return any surface water to the sewers.

There are some immediate advantages to a SuDS approach: for instance, porous paving can reduce accidents due to ice and eliminating blockable drains and gullies means fewer puddles on roads and pedestrian areas. There are wider-reaching benefits, which include a reduced risk of flooding generally, improved water quality in rivers and streams, more environmental areas and wildlife habitats and replenishment of groundwater supplies.

Summary: sustainable surface water disposal

Adopt a Sustainable Urban Drainage Systems approach to storm-water drainage and provide green roofs.

Sewage treatment

This section suggests that it is often best to connect to the mains drainage system if available. It goes on to outline the options for on-site treatment if the site is not served by mains drainage and also outlines radical alternatives which can be used.

Mains drainage

About 96% of the population is connected to a main drainage system. This carries sewage to a treatment plant which transforms it into more or less clean water and sludge. It is generally the safe, convenient, cost-effective and least energy-intensive option. The process is monitored and the standards required for discharges of effluent are rising under the influence of, among other things, the EU Urban Waste Water Treatment Directive. It is often best to connect to the main drainage system if it is available, and you may be obliged to do so if the site is served by mains drainage. However, one alternative is to treat the sewerage on site and return the cleaned water to the soil or into a watercourse. This is fine if the system is designed, constructed, monitored and maintained correctly.

On-site treatment

Of the 4% of the population not served by mains drainage, most use a septic tank and soakaway that disposes of the tank effluent into the soil. When this is not possible due to soil conditions or lack of space, there are several methods of making the waste-water clean enough to discharge to a watercourse. The appropriate system is the one that causes the least environmental impact when viewed from a lifecycle perspective. It may not be the solution that is apparently the most 'green'.

There are many techniques for on-site waste-water management. It is important to understand their various merits and drawbacks so as to pick the most appropriate technique for a particular circumstance. There is a risk that systems may not perform to the required ecological standards, which will be a problem for both residents and the environment.

Appropriate solutions are very site-specific. If a small volume of effluent is to be discharged to a large river (or to the ground), the level of appropriate treatment will be far less stringent than when a large volume of effluent is to be discharged to a small stream or close to a groundwater supply. Higher levels of treatment than appropriate usually incur environmental (and financial) costs that may offset the assumed ecological benefit of a purer effluent.

Most systems rely on a primary treatment stage of physical settlement to remove gross matter followed

FACING PAGE The sewerage from this row of houses is treated by reeds at one end of the pond.

by a secondary biological treatment using aerobic micro-organisms. This can sometimes be followed by a tertiary stage that further clarifies the effluent and removes nutrients and pathogens. The following options can be designed in different combinations to suit the particular circumstances and use.

Primary treatment options include:

- *septic tank* in which the solids and up to one half of the organic load are retained as crust or sludge. This is the most common installation, it is simple and economical, but requires the sludge to be emptied (usually every couple of years)

- *packaged units* generally include secondary treatment. They are a compact, readily available, medium-cost alternative with minimum maintenance often carried out by the installer under contract. They do use power, and no treatment is possible if there is power or mechanical failure. There are several alternative types including the Rotary Biological Contactor (RBC) which has a series of large-surface-area plastic discs half submerged in the effluent. They rotate slowly, powered by a motor, so exposing to air a film of micro-organisms that builds up on the discs; this creates the conditions for aerobic treatment. Another type is the Sequencing Batch Reactor (SBR) which aerates a batch of incoming sewage with air in a single chamber, then lets the sludge settle out and finishes the sequence by drawing off the effluent.

Secondary treatment options include:

- *Percolating filter* This is an open tank filled with clinker or stones. The effluent passes over these, forming a thin film. This exposes to air the

micro-organisms flourishing in it, allowing aerobic treatment. Clumps of micro-organisms must then be cleared from the effluent in a settlement tank. Circular versions with a rotating arm to distribute the effluent are a common sight in small sewage-treatment works. This is a tried and tested technique that requires little power, although it is relatively expensive. It is not generally appropriate for a small number of homes.

- *Reedbeds* A vertical-flow reedbed is similar to a percolating filter but with a layer of sand on top, planted usually with common reeds. A following settlement tank or second reedbed are usually needed. Reedbeds achieve a high level of treatment without the use of power, and maintenance is straightforward but does require awareness of the processes. It needs more space than some other methods, around 12m² per family home, and is relatively costly. The specification of the sand and the size of the bed are critical to prevent blockage. A horizontal-flow reedbed is shallower and designed to be full of water, unlike the vertical-flow type which is free-draining. The effect is to reduce the amount of oxygen, creating ideal conditions for the removal of nitrogen from treated effluent. A horizontal-flow reedbed is therefore best used at a tertiary stage, where it also removes pathogens. It can be about half the size of a vertical-flow bed, which is used for secondary treatment. Opinions vary on the role played by the reeds, but reedbeds almost certainly are not the single answer to sustainable sewage treatment, despite their iconic status.

- *Ponds* can be used for primary, secondary and tertiary treatment. They are usually seeded with aquatic plants and can look beautiful. A large surface area is required, around 50 to 100m² for a family house, and are relatively expensive. Aerated ponds require less area and should be odour-free, but typically use more electrical energy than package plants such as RBCs.

- *Leachfields* These commonly follow a septic-tank primary stage. A leachfield is a series of perforated pipes in trenches surrounded by gravel. This can be completely unseen below a lawn. It provides good-quality treatment at low cost and low maintenance. It is natural treatment in its simplest and most efficient form and is the preferred option for smaller on-site sewage-treatment systems where the soil is sufficiently free-draining to allow effluent to drain into the soil.

- *Living machines* is the name coined for an artificial ecosystem created with planting in a glasshouse and intended to imitate nature in the treatment of effluent. However, living machines are expensive to build and operate. They require power for pumps and aeration, need close management and operate at low efficiency. They are not 'green' from the viewpoint of environmental impact.

Composting toilets

The systems so far described all mix two useful resources – clean water and human manure – to create sewage. Sewage must then be treated to obtain water that is safe to release into the environment, and sludge which has to be disposed of safely.

The alternative is composting toilets. These use little or no water, don't get blocked, make no noise and, if designed well, no smell. They can be used with a urine-separating toilet bowl. This has a separate small bowl at the front for urine, which contains the majority of the nutrients in human waste for plants. Urine is sterile, so it can be mixed with other organic waste such as straw to make compost for food crops. A compost toilet will produce a small volume of surprisingly innocuous humus which can be used on trees and shrubs. Users will need to understand how the system works and use it responsibly. It will need emptying once a year or so depending on the design. Compact models have proved problematic and so larger systems are recommended, but these require a void or basement below the toilet bowl to house the composting chamber, and a good-quality model can be expensive. Wastewater from sinks, basins, showers and baths can be used for irrigating the garden.

Regulation

The planning authority will wish to be assured that your proposal for sewage disposal will not cause a nuisance. They may consult the local environmental health department if the technology is unfamiliar and they may also consult the Environment Agency (EA) – in Scotland, the Scottish Environment Protection Agency (SEPA) – who are responsible for monitoring pollution and the environmental quality of watercourses. If you wish to discharge into a watercourse, you will have to obtain Agency consent. This will specify the quality that your effluent will have to achieve. Single remote dwellings discharging to ground may not need formal consent but the EA or SEPA should be informed anyway and will advise whether permission is required. Building Control will also have to give drainage consent for whatever you propose. You will also want to be assured that your system is working properly and not causing pollution, so you will need to be familiar with the standards and how to test for them.

Costs and benefits

On-site treatment reduces the load on the main sewer system and may reduce sewerage charges. Composting systems additionally save water and enable nutrients to be returned to the soil and will reduce water costs. Capital costs depend on the type of system. On-site disposal systems require monitoring and maintenance.

Sustainable sewerage disposal

If mains sewerage is available it is often preferable, and you may be obliged to connect to it. If on-site treatment is necessary because there is no mains drainage or if you prefer to be self-sufficient, the appropriate system is the one that causes the least environmental impact when viewed from a life-cycle perspective. It may not be the alternative that appears 'greenest'. A composting system is the greenest alternative.

Environmentally preferred forms of construction

This section suggests forms of construction and materials for each element of the building which are preferred from an environmental point of view.

There is not a single best way to build

There is no one ecologically best way to build. A timber building can be highly insulated relatively easily, whilst a brick one will make better use of passive solar gains; timber is a renewable resource whereas bricks are a reusable one. What is important is to match the decisions on materials and construction to the particular circumstances and to make sure that there is an appropriate balance between different aspects of the project; to make sure that there is a high level of insulation which will reduce emissions for a minimum cash outlay, in preference to fitting an array of photovoltaic cells which will generate a small amount of power at considerable capital cost for example.

The Environmental Preference Method

This section is based on information from the Environmental Preference Method published in the Netherlands in 1993 and now used by the majority of local authorities there as the basis for the application of the principles of sustainable building in practice. Some of the information is related to Dutch building practice; the recommendations given below have been related to experience of UK practice and comparative assessments of environmental impact contained within *The Green Guide to Housing Specification* published by the Building Research Establishment. The *Green Guide* does have one significant limitation, which is that it assesses complete composite elements of construction – walls and roofs, for example – and it is not possible to isolate the performance of individual parts of the element, the insulation or cladding for example.

The Environmental Preference Method compares materials and products currently on the market and ranks them according to their environmental impacts. It is based on currently available data, and is better than Life Cycle Analysis for making choices in construction and materials. The information given is applicable to elements of construction for both new building and refurbishment. The method does not include considerations of cost or aesthetics – it is an environmental preference. However, a suggested preference is given based on a low environmental impact solution which is proven in practice and which is of no or negligible extra cost compared with currently available alternatives.

The life-cycle of materials

The Environmental Preference Method assesses impacts throughout the whole life-cycle of a material or product:

* *Extraction* – the rate of depletion of stocks of resources together with an assessment of the damage to natural ecosystems resulting from extraction processes, for example from mining for bauxite, harmful emissions from coal mining and the risk of environmental disaster from the transportation of oil.

* *Production* – processing raw materials into products results in harmful emissions to soil, air and water, as well as the creation of waste and energy consumption. In general, the more highly manufactured a product is the more the environment will be threatened.

* *Building* – the principal environmental problems arising during the building phase are the consumption of energy and the creation of waste and pollution from dust, noise and vibration. Also, the care taken during construction influences the lifespan of the elements of the building and the lifespan of the building as a whole.

- *Occupation* – the impact of materials choices on the health of occupants, arising from noxious emissions from chipboard, paints and adhesives, for instance. Also problems arising from poor indoor environment from noise, draught, damp. Also energy use, emissions, waste and the effects of on-going maintenance.

- *Decomposition* – Demolition creates a large amount of waste with emissions to air from incineration, and water and soil from landfill. The reuse of components with a minimum of reprocessing is preferred to recycling materials to create new raw material.

The environmental issues considered

The main issues considered to ascertain the environmental preferences are:

- *Damage to ecosystems*

- *Scarcity of resources*

- *Emissions* – noxious substances released during the course of the life cycle of a material or product. An example is heavy metals from timber preservatives. Also disasters can occur: although the risk may be small, the consequences could be catastrophic.

- *Energy use* – during extraction, production and transport. This implies the use of scarce resources on the one hand together with the release of harmful emissions which contribute to the greenhouse effect, acid rain and smog production on the other.

- *Waste* – can cause numerous problems including difficulties in separation. Poor degradability, dust, leaching into watercourses from landfill, and the release of noxious substances on incineration.

- *Reuse* – can be promoted by using materials that can be reused after demolition without reprocessing. This reduces waste and the use of resources.

- *Lifespan and reparability* – greater durability reduces the need for repairs and replacement. The durability of components needs to be considered in relation to the life of the building as a whole.

A sustainable strategy for selecting materials

The basic strategy for the selection of sustainable building materials consists of the following steps:

- *Efficient use of materials* – evaluate the need for a new building and consider refurbishment, design the plan and construction for the efficient use of materials, optimise the size of components and match the lifetime of components with other elements of the building.

- *Use renewable and recycled resources* – avoid composite materials that cannot be separated at the end of their life, avoid gluing and sealing components together and design for dismantling rather than demolition.

- *Use materials with low environmental impact* – use unprocessed materials such as solid timber, natural stone or earth, or materials with low embodied energy, locally sourced to minimize emissions from transport. Avoid the use of high-energy heavily manufactured products as far as possible. Select materials to create a healthy internal environment and to minimize the use of energy through high levels of insulation, airtight construction, controlled ventilation and so on.

Each element of the building is considered in turn, and first, second and often third preferences are given for the construction of that element together with materials that are not recommended on environmental grounds.

Foundations

Foundations are required to carry the loads of the house without movement causing cracking and rendering the building unusable. Generally, the sustainable approach is to aim for the minimum use of resources: steel, which is relatively energy-intensive; cement, which is one of the most significant sources of the emissions which lead to climate change; and aggregate, which is in short supply. Recycled aggregates should be used where possible, although current practice limits this to 20% of the total to retain the strength characteristics of the concrete. In this connection, note that recycled aggregate can be transported for 150 miles before the environmental impact of transport exceeds that of winning virgin aggregate. Extensive foundation construction also generates large quantities of spoil. It is preferable, but not always possible, to dispose of spoil on site rather than exporting it to a landfill. This eliminates the cost and emissions resulting from transport and avoids the cost of Landfill Tax.

Traditionally, stone and timber buildings have been founded on strips of shallow stone. Later, the walls of brick buildings were widened below ground level to form a shallow spread footing.

These days, conventional load-bearing masonry construction requires continuous strip foundations under the walls. Generally, foundations need to be at least 900mm deep to avoid the effects of frost heave. They may, however, need to be deeper in certain circumstances:

- *If there are trees close by*, as these can cause movement both from their roots but also from the fact that they draw moisture from the soil which can cause some types of clay soil to shrink. In this case, the depth may be up to 3m, depending on the moisture demand for the type of tree, its proximity to the building and the shrinkability of the type of soil.

- *Bad ground which is not stable enough* to support the building adequately. This may be caused by very wet ground or loose fill which may include debris from earlier development, a disused waste pit, or mining subsidence.

If the depth is required to be in excess of 2.5m, it is generally the case that it is cheaper and easier to use piles with ground beams between to support the walls. An alternative to this is a raft foundation which spreads the loads. This has to be a very substantial slab of concrete, with a great deal of steel reinforcement to carry the loads. This method is both expensive and uses a great deal of resources in its construction.

A frame construction with non-loadbearing infill can be supported on isolated rather than continuous foundations. This is both economical and resource-efficient, as it reduces the amount of concrete to about 20% of a continuous foundation. You will have to avoid heavy brick cladding, which is vulnerable to cracking if there is any movement in the foundations. Load-bearing timber-frame panel construction can be supported in this way provided edge beams to the ground floor are designed to take the loads from the wall panels.

These isolated foundations can be either pad foundations – consisting of a block of concrete around 600mm square to support the loads of a typical house and 900mm deep, which can be excavated by hand – or circular holes can be drilled out using a machine-mounted auger. A post-hole auger machine, either self-powered or mounted on the back of a tractor, can be hired to drill holes up to 600mm in diameter and up to 1200mm deep. If the area of a 600mm diameter pad is insufficient to support the load, the bottom of the hole can be under-cut by hand to increase the bearing area and thus the bearing capacity of the pad.

If the isolated foundations need to be over around 1.5m deep, you will need to use piles. This requires specialized machinery, and will generally be installed

FACING PAGE Above: A piled foundation without concrete ground beams can be a quick and economical solution for timber construction. Timber suspended floor beams span between galvanized steel shoes bolted down to the pile caps. Below: The ground floor can be supported – here the concrete was cast in a plastic flowerpot.

by a piling subcontractor, although it is possible to hire mini-piling equipment. Conventionally, piles are of concrete. A piling rig makes a hole around 200mm in diameter and a steel cage of reinforcement is placed in the hole, which is filled with concrete. A proportion of recycled aggregate reduces the environmental impact. Better is using timber, a renewable resource. This is common in marine construction, and is used for buildings in the Netherlands. In the UK, the advice is that piles should be of hardwood. Dutch practice is to have a concrete pile cap at ground level. These can be pre-cast.

Conventionally, the piles are tied together at ground level with reinforced concrete ground beams. This is a relatively complex and expensive operation. It is possible to avoid ground beams by designing a suspended timber ground floor which is capable of resisting any tendency for the piles to move sideways. The floor will have to be bolted to the base using galvanized steel shoes which can transfer any horizontal loads from the piles to the floor construction.

Conventional masonry construction requires a level plateau for the house to be built on. On a sloping site, this implies extensive earth-moving, spoil and disruption to the natural slope, groundwater and appearance of the site. Building above the ground rather than on it eliminates the need to level the ground, with the attendant savings in cost and disturbance to the natural site. A useful undercroft can be created. Isolated bases can be set at the natural ground level without levelling the site. This eliminates heavy earth-moving, which can be a benefit for self-builders in particular. The positions of the bases are not absolutely critical either, which is also an advantage at the stage when the site is largely mud with few fixed points.

The timber frame can stand on the bases: directly on a pre-cast concrete paving slab, or raised above the splash zone on a concrete stool made by filling a plastic bucket with concrete set on a mortar bed. Ground floor beams can also be supported on short

FOUNDATIONS				
First preference	**Second preference**	**Third preference**	**Not recommended**	**Comments**
Timber piles Concrete piles with recycled aggregate and no ground beams. Pad foundations with recycled aggregate House raised above the ground on posts to avoid earth-moving.	Concrete strip footing	Concrete piles with ground beams. Concrete trench fill Concrete raft foundation		Timber piling is not common practice in UK. Recycled aggregate may not be available or of untested strength. Not suitable for masonry construction.

concrete columns reinforced with steel rods cast on top of the bases in a permanent shuttering of plastic drainpipe. In all cases a polythene damp-proof membrane is required between the foundation and the timber. This can be obtained with a bitumen layer applied to it which can seal the vulnerable end grain of timber posts.

When designing and building foundations, avoid the use of reinforced concrete as far as possible as it is resource-intensive and difficult to recycle when no longer useful. Use recycled rubble for fill and polythene for damp-proofing.

Basement construction tends to be relatively massive to resist the lateral loads from the soil, and consequently has a relatively high environmental impact; although a basement can create useful accommodation within deep foundations at reduced extra cost. Timber is used for basement construction in North America.

Ground floors

Traditionally, floors were constructed of stone slabs or clay tiles bedded on sand. Damp was kept out by providing good ground drainage around the building. Alternatively, timber floors spanned between the walls above the ground. Joists were built into the walls, which were often damp, which caused the timber to rot. Good ventilation to the under-floor void is necessary – but not too much as

Building above the ground rather than on it saves the cost and disruption caused by earth-moving. It can reduce costs and also creates a useful undercroft for storage.

GROUND FLOORS				
First preference	**Second preference**	**Third preference**	**Not recommended**	**Comments**
Timber	Beam & block flooring with hollow concrete blocks.	Hollow pre-cast concrete planks for party floors between flats	Solid concrete	Timber floor must be airtight and protected from moisture Recycled aggregate is preferred if available Hollow clay blocks are preferred if available.

it can lead to excessive draughts through the floor, leading to high heat loss.

A suspended timber floor uses renewable resources and is relatively easy to insulate to a high level, using low-environmental-impact insulation materials such as cellulose fibre; but a timber floor must be made airtight, by either fixing tongue-and-groove (T&G) flooring panels with glue or including an airtight membrane in the construction. Joists can be supported on brick walls using steel hangars to protect them from moisture. Timber boarding or FSC-certified OSB (oriented strand board) are preferred to uncertified OSB, chipboard and plywood as the floor deck. Services can be incorporated relatively easily, but care has to taken to achieve either a level or ramped access for people with limited mobility.

Timber floor panels can be prefabricated, which reduces site work and is good from an environmental point of view as it tends to reduce waste; but prefabrication tends to increase costs. Composite timber I-beams can be used in place of solid timber sections, which reduces heat loss through the cold bridge formed by solid timber. Composite beams use a thin web of either OSB or even better Masonite, which is a form of structural hardboard made from sawdust.

Left: An alternative form of engineered timber beam consists of joists made with steel struts between softwood top and bottom flanges. The open webs allow services, especially ventilation ducts, to be accommodated easily. Right: I-beams are light and strong and eliminate thermal bridging. They are good from an environmental point of view because they are a very efficient use of timber.

This is environmentally good because it is bonded using the natural cellulose in the timber unlike OSB, which relies on synthetic resin. The flanges are either softwood or stronger parallam (which is laminated timber similar to plywood but in larger sections and lengths similar to ordinary timber). An alternative form of composite joist is made from an open web of lightweight galvanized steel struts between small softwood flanges at the top and bottom. Services, particularly large ventilation ducts, are easily incorporated through the open webs of the joists. These composite beams use a minimum of timber which is also in small sections. They are strong, lightweight and dimensionally stable, but they cost more than solid sections for most purposes.

Beam and block flooring is relatively efficient in the use of materials, and can have some thermal insulation value if you use lightweight concrete blocks. Its environmental performance is enhanced if you use hollow blocks and if you can use a proportion of recycled material in the concrete. Clay blocks are a better solution environmentally as the raw material is more plentiful, but clay blocks are not commonly available in the UK. Polystyrene blocks are available to improve the thermal performance, but these are a petrochemical-based product with the environmental implications that follow. The pre-stressed concrete beams which support the blocks are manufactured with an upward camber which is designed to create a level floor without deflection when loaded with heavy concrete blocks. When lightweight polystyrene blocks are used the camber does not flatten out and can cause difficulties in achieving a level floor.

A solid concrete floor may be required (in a garage or workshop, for instance) but it is much heavier and less efficient in its use of materials. Services are more difficult to incorporate, and insulation materials that can carry load are required, which tend to be more expensive and have greater environmental impact.

The acoustic performance of timber intermediate floors within a dwelling can be improved by including cellulose fibre between the joists and increasing the weight of the construction by adding an extra layer of plasterboard to the underside. A floating floor with the floor deck not directly fixed to the structure is required to party floors between flats to prevent impact sound passing through the floor, and plasterboard provides a fire separation.

Floor insulation

Air-based insulations use different materials as the matrix to encapsulate air pockets. These materials can either be of:

- *Organic origin* from natural vegetable matter that is both renewable and recoverable on demolition. These materials generally require very little energy in their production. Examples include cork, wood fibreboards, hemp, sheep's wool and loose cellulose fibre from recycled newspaper,

- *Inorganic origin* from naturally occurring minerals which are generally not renewable but plentiful, generally involving moderately high energy inputs and consequent emissions. Examples include mineral wool, fibreglass, vermiculite, foamed glass and aerated concrete, or

- *Fossil-fuel origin* which is not renewable, with generally high energy inputs, emissions and pollution implications from the chemical industry. Examples include expanded and extruded polystyrene, foam polyurethane and foam polyisocyanurate. Ozone-depleting gases have now been eliminated from the manufacture of foam materials.

The environmental preference is for insulation materials from natural organic sources. Fossil-fuel-based materials should be avoided if possible. Insulation containing ozone-depleting gases should be avoided at all costs.

Cellulose fibre and sheep's wool are the preferred materials, as they are renewable. Neither will sup-

FLOOR INSULATION				
First preference	**Second preference**	**Third preference**	**Not recommended**	**Comments**
Cellulose fibre Sheep's wool	Mineral wool Expanded polystyrene	Foamed glass	Polyurethane Extruded polystyrene	Avoid ozone-depleting products

port load, and they require a void to be installed in, for example as in a timber floor. A floating timber floor can be constructed above a solid floor to accommodate renewable insulation. Cellulose fibre insulation is produced from recycled paper, and wool is virtually a waste product from the production of meat. Cellulose fibre, however, has to be installed by a specialist contractor using a blowing machine. This means that it is less flexible to use because you have to have all the construction which needs to be insulated ready at one time, whereas it is often convenient to insulate some parts – say the roof – in advance of constructing the internal partitions, for example. However, it does mean that the insulation fully fills the voids into which it is injected, unlike

quilt materials which are often imperfectly cut into the spaces between wall studs or rafters.

Sheep's wool is much more pleasant to handle than other similar materials such as mineral wool, and does not release fibres which cause irritation when inhaled or when they come into contact with your skin. Sheep's wool also has the property of being resilient, such that it will stay in place in an overhead cavity when compressed slightly and sprung into place between roof joists for example. This makes it much easier to use than mineral wool, which will just fall out of an open cavity. However, it is significantly more expensive than mineral wool.

Other insulation materials are available in slab form, and selecting the correct density will allow them to carry load and be installed below a concrete slab. This will increase the thermal capacity of the construction – that is, the amount of heat that is required to raise the temperature of the construction. It is a measure of the capacity of the building to store heat and is a function of the material (concrete has a higher thermal capacity than timber, for instance) and the quantity of material in the building. Insulating below the concrete slab includes the slab in the thermal capacity of the building, and is a requirement for under-floor heating with a solid floor. Edge insulation is required to prevent cold bridging at the perimeter of a solid floor.

Expanded polystyrene is the least damaging fossil-fuel-based insulation material; extruded polystyrene is much worse, but is a closed-cell product which does not absorb moisture and is therefore useful as

Cellulose fibre insulation made from recycled newspaper.

WALLS: STRUCTURE				
First preference	**Second preference**	**Third preference**	**Not recommended**	**Comments**
Timber-frame Earth Straw bale Local stone Recycled bricks	Sand-lime bricks	Hollow concrete blocks with recycled aggregate SIPS	Concrete blocks with no recycled aggregate New bricks Cement mortar	Timber to be untreated Concrete blocks to include recycled aggregate Use lime mortar

upside-down roof insulation and in areas which are at risk of condensation.

Polyurethane is the most efficient of the materials listed but is the most environmentally damaging.

Walls - Structure

Frames

Frame structures are preferable to masonry. Many old buildings were constructed using heavy oak timber-frame construction, but often the timber was 'green' or unseasoned, leading to shrinkage and movement as the timber dried out, and we have an inheritance of quaint irregular timber buildings as a consequence. This has not been a problem as the frame was generally infilled with relatively flexible wattle and daub, or brick set in lime, which can accept that level of movement without cracking. This type of infill between the frame, which shows as a black painted frame with white limewashed infill, does not keep out driving rain, and so the frame has often perished. Buildings clad with timber weatherboarding have tended to keep the rain out and have lasted better.

Generally, the carpentered joints of traditional framing have been replaced with bolts, or nailed stud panels replace frames and heavy green oak has

been replaced by calculated, lighter sections of kiln-dried, structurally graded softwood. However, there has been a resurgence of interest in traditional heavy green oak framing in recent years, prompted by the desire by some to reflect traditional values. Roundwood can be used, which uses the tree in its natural state with the bark removed. This can be done mechanically to produce a telegraph pole type of member, or the bark can be stripped by hand which retains the natural form and texture of the tree. Either way, the fibres in the tree remain intact and the member is stronger as a consequence.

Engineered timber

So-called 'engineered' timber sections have been developed, which take elements of timber and combine them to form composite members which are lighter and stronger than solid timber and which are more dimensionally stable. Composites make use of poor quality timber and small sections and waste timber, increasing the proportion of the timber harvest which can be used. First to be developed was laminated timber, which glues strips of timber together in layers horizontally to make a member of consistent strength without defects. This has been developed further in Laminated Veneer Lumber (LVL), which is manufactured from thin veneers glued together vertically to make what is in effect a beam made of plywood. Members are made in two versions. One has some cross veneers as in plywood, and is suited for taking compression loads as well as the bending loads applied to a beam. This makes it very dimensionally stable and it is used for rim-beams within the wall construction at the perimeter of a timber floor in timber-panel construction. In the other version, all the veneers are orientated in the same direction to give maximum strength for carrying the bending loads, as in a beam.

Another form of engineered timber is Parallel Strand Lumber (PSL), often known as Parallam. This is made from oriented strands of timber rather than veneers as in LVL, and is therefore related to OSB (Oriented Strand Board), but in the form of a timber member rather than a board. It is not as strong as LVL, but it has a pleasant texture which enables it to be used to achieve a fine finished appearance.

For carrying lighter loads as in joists or rafters, I-beams have been developed with a thin web connecting more substantial flanges at the top and the bottom. This places the material where it is most effective for load carrying (at the extremities), and is of similar shape to steel beams. I-beams reduce cold bridging to a minimum – that is, the effect of an element of construction with poor thermal insulation forming a 'bridge' between the outside and inside faces of an element of construction, such as a wall allowing heat to pass from inside to outside, thus increasing heat loss. This also forms a cold spot, which will also be at risk of condensation forming. The light web can be OSB or Masonite, a structural

grade hardboard made from compressing sawdust under high pressure and temperature. It is environmentally good because it relies on the natural cellulose in the timber to bind it together rather than manufactured adhesive. The flanges can be softwood or LVL can be used for greater strength.

Another engineered timber beam, called Eco Joist, uses light steel struts to connect the flanges. This permits large-diameter ducts, such as are necessary for a whole house ventilation system, to pass through the floor construction without difficulty. Alternatively, it is possible to make your own economical, composite members on site for use as studs or rafters from softwood flanges nailed together with ply patches.

Materials for frames

Frame structures are light in weight, can be highly insulated relatively easily, and the non-loadbearing infill can be designed in a flexible manner and can be changed at any time in the future. Timber is the preferred material. It is a renewable resource and is relatively easy to work using simple tools. The timber should not be treated, and so should be designed to avoid rising damp, moisture penetration from outside and condensation within the construction. Steel framing is also good. Although it takes a relatively large amount of energy to manufacture, it contains a high proportion of recycled material; it is strong, which leads to the efficient use of material; and it is very adaptable as it can be fabricated and jointed using bolts or welding in any shape you wish. This is unlike timber, which has very different properties along the grain to across the grain: timber has high tensile and compressive strength along the grain – in other words it will withstand pulling or pushing along its length – whereas it will split or be crushed across its length, which also limits the design of joints. Steel is generally (and timber is often) prefabricated into panels off-site. This reduces site work, and is good from an environmental point of view as it tends to reduce waste, but prefabrication

The posts and beams of this house are made from Parallel Strand Lumber, PSL, also known as Parallam. Made from strands of wood bonded together under pressure, it is strong and dimensionally stable.

Structural Insulated Panel Systems (SIPS) are made from foam plastic insulation sandwiched between OSB boards. They can be used for walls and roof to quickly create a well-insulated, airtight envelope.

Another lightweight form of structure is SIPS (Structural Insulated Panels), formed by bonding ply or OSB to either side of an insulation board core, commonly expanded polystyrene, to form a sandwich panel. These are used for the roof and walls to form a very well insulated and airtight shell.

Masonry

Masonry construction can be made of stone, unfired earth or brick. Stone should be sourced as near to the site as possible, and traditionally was used in Wales and Scotland and in the areas of England where suitable building stone, including limestone, granite and sandstone, occurred near the surface in bands running across the country – which can be detected in the colour of stone buildings that remain. Second-hand material is preferable to new stone or brick. Sand-lime bricks are better than hollow concrete blocks in terms of the reduced exploitation of raw materials, but concrete blocks can be made to provide some thermal insulation. Lime mortar enables stone and bricks to be reused when demolished.

Earth building

Earth construction is the best from an environmental point of view. In the past it involved little or no energy inputs other than manpower, and even now, although machines are used for the bulk moving and placing of soil, it is still the form of building which has by far the lowest embodied energy. It is a very low-cost way of building, and offers high thermal mass, which makes it particularly appropriate for passive solar architecture. It avoids potentially toxic substances and returns to the earth when no longer useful.

Earth suitable for building is available on over two-thirds of the globe's landmass, and is used to house one-third of the world's population. The Romans brought earth building to France and the Moors brought it to Spain, from where it was

increases costs. Steel frames using heavy so-called hot-rolled sections are widely used in commercial construction, but smaller-scale buildings generally use lighter cold-rolled sections which are formed by passing unheated sheet steel through a rolling mill to form angle and U-shaped channel shapes. These can be drilled and screwed together using hand-held power tools. Reinforced concrete frames are not good environmentally: they are a high-energy, high-emissions form of construction that is difficult to dispose of when demolished.

exported to Mexico and the south-west of North America. Earth building was established in Australia in the middle of the 19th century, and there has been a recent resurgence in earth self-build there. There has been a continuing tradition of earth building in central Europe and Britain in the East Midlands, East Anglia, West Wales and most particularly in Devon. There was a revival of interest between the wars, and more recently eco-builders have been building a number of earth homes.

There are a number of methods of building with earth which have developed in different places to suit particular circumstances which are outlined below. However, there are a number of characteristics which are common, which include the need to ensure that the earth construction is protected from rain. Foundations must protect the walls from rising damp and water splashing up the wall at the base. Good drainage around the building and a weatherproof zone at the base of the wall are important. In the UK climate, a deep overhang to the roof is necessary to protect the walls. The composition of the soil is important: a minimum proportion of clay is required, an even distribution of particle size from fine silt through sand to small gravel is desirable, the moisture content is critical, and organic matter must be avoided. Straw can be incorporated to reduce weight and reduce cracking, cement can be added to enhance strength, and asphalt emulsion can be added to stabilize the blocks and provide water repellance, although adding stabilizers reduces its environmental performance. Earth has very little insulation value, and additional insulation is generally incorporated.

Adobe

Adobe is mud-brick construction – known as 'clay lump' in Britain, where it was used in East Anglia during the 18th century. The oldest permanent dwellings that have been excavated, in the Indus Valley dating from 8300 BC, are of mud-brick construction. Then, as now, mud is moulded into

Mud-brick or adobe construction is still common in some parts of the world, using local, low-cost, abundant resources which will return to the earth when no longer needed. Earth buildings have a high thermal capacity, which is useful for passive solar heating.

bricks which are dried, bonded together with mud mortar, and rendered with mud plaster. Adobe is a relatively quick way of building with earth, as compared with cob or rammed earth. It is also possible to construct vaults using mud bricks. The so-called Nubian vault can be constructed without formwork. A gable wall is constructed against which an arch is built at a slight angle. Successive arches are built leaning against the previous arch, which supports it without centring. Mud brick construction has been developed by the introduction of the CINVA ram, which applies high pressure to produce a compressed soil block.

Cob

In cob construction, mud is applied to walls in layers around 450mm thick and shaped by hand without formwork, leading to a somewhat sculptural appearance. The mix takes around two weeks to dry, at which time it can be trimmed to form a slightly tapering wall around 600mm thick which is then generally plastered with a lime render. Cob is a very simple but rather slow method of construction. The mix can be improved by the addition of long straw and coarse sand. Cob has been the most common form of earth building in Britain.

Rammed earth

Also known by the French term *pisé*, this consists of soil rammed into high strength formwork in layers. This can produce a very precise finished product, which is often in the form of rectangular panels. It is a labour-intensive process, which is often mechanized these days using a digger to raise the soil and pneumatic rammers for compacting the soil. Rammed earth has greater load-bearing capacity than other forms of earth construction.

Modular contained earth

This is an approach to rammed-earth construction which relies on temporary formwork in the form of either motor tyres or 'earthbags' which are filled in situ on the wall and tamped into place and rendered with either mud or lime plaster. The bags can be either hessian or polypropylene long 'sausages' or short bags. The tyres take three or four wheelbarrows of soil, reusing tyres which are otherwise a huge waste problem, with thousands being dumped in landfill every year. A series of passive solar 'Earthships' have been built in the south-west of the USA, and one in Brighton, which consist of a U-shaped wall of tyres filled with earth supporting an earth berm with a green roof and windows facing the sun.

Light clay

Light clay or *Leichtlehm* is a composite material invented in Germany in the 1920s as a development of wattle and daub infill. It consists of straw, reed or wood chip coated with a slurry of clay. This is then packed into forms to make walls, blocks or panels which are used for non-loadbearing infill. The clay binder preserves the organic fibres and provides protection from fire and rodents. The material is half as dense as adobe, and has some insulation value. A variant is to use hemp in combination with gypsum or lime. This has recently been used to construct a number of homes in East Anglia, as described in one of the case studies (see page 91).

Straw bale

Straw is used as reinforcement in earth construction, but was used as a structural material in its own right with the invention of the mechanical baler. At the end of the 19th century bales were stacked to form load-bearing walls on the treeless plains of Nebraska, the so-called 'Nebraska method'. Alternatively, straw can be used as infill within a timber-frame structure. A number of the early straw-bale buildings have survived over a hundred years; they engendered a revival of interest in the 1970s and 80s, and the building technique became popular in the 1990s. It is a very cheap material with very low environmental impact, and makes use of a waste by-product which is no longer permitted to be burned. It is biodegradable when no longer needed, and has moderate insulation properties. Straw bales are too dense to burn or to be attacked by rodents. It is a building technique which does not require close tolerances, and is particularly suited to rough plaster finishes.

Stacking up straw bales as the structure, and plastering them inside and out with a lime render to create a sandwich wall with stress skins on each face is a very simple method, but there are complications because the bales settle in time under load and cause cracking of the render, which in turn can lead to rain penetration. A rainscreen cladding of timber weatherboarding or some other material can be used in place of a rendered finish, as described in one of the case studies. Like earth construction, it is vital that the straw remains dry at all times, including during

A wall of a hemp and lime mix within a timber frame.

WALLS: CLADDING				
First preference	**Second preference**	**Third preference**	**Not recommended**	**Comments**
Durable timber Shingles	Brickwork Tile hanging Lime render	Fibre/cement, resin-bonded and plywood cladding panels	Non-sustainable tropical hardwood Treated softwood Cement render	European timber is preferred to Canadian cedar. Reclaimed bricks and lime mortar are both good.

construction, to prevent it rotting; and so correct detailing at the base and a deep roof overhang are important. Arrangements to pre-compress the bales can be devised, using threaded steel rods and timber. Alternatively, the straw can be used as non-load-bearing infill to a timber frame, and can be used for two-storey construction. Movement in the infill can be accommodated, and the straw can be stacked up under the protection of the roof.

Straw has similar insulation properties to other materials, but straw bales create a very thick wall: 600mm, or slightly thinner at 360mm if laid on edge in a non-structural infill to create a very low-cost super-insulated wall and a wall with substantial thermal mass. One of the effects of the thick walls is that windows have deep reveals – this is useful as a deep window sill or window seat, but it does mean that the inside of the building is somewhat remote from the outside and the amount of daylight can be reduced.

Walls – cladding

Brick and stone can be fairfaced, or be more cheaply constructed and then painted or rendered. Earth walls are generally rendered with lime or mud plaster.

Left: Douglas fir is classified as a moderately durable species, here used as cladding without any applied finish to eliminate maintenance. Middle: Tile hanging is another lightweight, maintenance-free cladding. Right: Lime render is preferred to cement render because it requires less energy to manufacture and is less susceptible to cracking.

Lightweight cladding panels are maintenance-free.

Walls - insulation

Cellulose fibre and sheep's wool are the preferred materials as they are renewable. Both can be used in a breathing construction which is an effective way of reducing the risk of condensation within the construction. Both are made from waste products – cellulose from recycled paper, and wool is a by-product of meat production. Cellulose has to be either wet-sprayed into place against a backing which requires access or (probably easier) placed into voids by a combination of vacuum and air pressure. Either way, specialist machinery is required and a subcontractor has to do the job. Cellulose fibre cannot be used in a masonry wall cavity and requires an enclosed void to be blown into. If installed correctly, cellulose fibre will fully fill voids, thus reducing the risk of cold spots arising from the imperfect fitting of fibreglass or mineral wool slabs or batts. Wool is pleasant to handle, unlike fibreglass or mineral wool which require protection for skin and mouth.

Expanded polystyrene is the least damaging fossil-fuel-based insulation material; extruded polystyrene is much worse, but is a closed-cell product which does not absorb moisture.

If you are refurbishing an existing house, you will be concerned to improve the insulation of the walls. This is not an entirely straightforward job, however, and there are some basic principles to bear in mind. If there is a cavity, you should fill it with insulation as this is the easiest and most cost-effective way of improving the performance of an existing wall. The preferred material is mineral wool injected into the cavity followed by expanded polystyrene. Cavity fill alone will not achieve a very high performance, and you should consider adding additional insulation. Insulation will also be required to bring a solid wall up to a reasonable thermal performance. This will have to be either inside the existing wall or outside. Inside insulation

Durable wood (that is, timber that does not require treatment or painting) is the preferred cladding material. It can be in the form of boards or shingles. Durable timbers include cedar, oak, sweet chestnut and Douglas fir and larch, specified to avoid sapwood, and detailed to avoid being permanently damp. Cedar is imported from the west coast of Canada, which involves substantial transport energy inputs, and so European alternatives are preferred.

Brickwork requires energy to fire the clay, but is relatively good because it requires little maintenance, and is preferred to the range of cladding panels available (which includes faced non-tropical plywood, fibre/cement and resin-bonded panels). Reclaimed bricks are good. However, brickwork has to be fully supported, which may increase the cost of foundations for lightweight construction. Also good are tile hanging and lime render. On a timber-frame building, this can be on a stainless-steel mesh backing with a ventilated cavity behind. Treated softwood, cement render and tropical hardwood from non-sustainable sources are not recommended.

WALLS: INSULATION				
First preference	Second preference	Third preference	Not recommended	Comments
Cellulose fibre Sheep's wool	Mineral wool Expanded polystyrene	Foamed glass	Polyurethane Extruded polystyrene	Avoid ozone-depleting products.

does not affect the appearance of the building, but it does reduce the size of rooms slightly. It also leads to a risk of cold bridging occurring where the external wall meets internal walls and floors, and it is generally more difficult to achieve high standards of insulation because there is a limit to the thickness of insulation that can be readily accommodated. External insulation can be carried out whilst the house is being lived in, but is generally more expensive. The preferred materials for both internal and external insulation are similar to the preferences for new construction. Particular care must be taken to prevent rising damp or damp penetrating the walls if you are intending to use cellulose fibre to insulate existing walls, as the insulation will degrade if it gets damp. Both cellulose fibre and wool require a cavity to be formed with battens in which to place the insulation. Cork is a renewable insulation material which avoids this problem, but it is relatively expensive.

Internal walls

Framed partitions using either steel or timber have less environmental impact than solid walls. They are not as effective against sound, however, and their performance can be improved by including cellulose fibre or mineral wool insulation in the cavity. Special fixings are needed to attach fittings to hollow walls. Earth walls are a low-environmental-impact solution, with excellent acoustic properties. For solid walls, sand-lime bricks are better than hollow concrete blocks in terms of the reduced exploitation of raw materials.

WALLS: INTERNAL WALLS				
First preference	Second preference	Third preference	Not recommended	Comments
Timber-frame Earth	Sand-lime bricks	Hollow concrete blocks		Timber to be untreated Concrete blocks to include recycled aggregate Use lime mortar

ROOF FORM				
First preference	**Second preference**	**Third preference**	**Not recommended**	**Comments**
Pitched roof	Flat roof			

Roofs

Form

A pitched roof naturally sheds water, and permits the use of concrete and clay tiles, which are less environmentally damaging than the membranes that are required to waterproof a flat roof. The form of a pitched roof, however, can limit the plan arrangement of the house, and it will tend to limit the options for extending the building later.

A flat roof is more economical and easier to build. They have a reputation for failures caused by leaks and condensation, but these can be avoided if designed with certain principles in mind. The minimum-risk design has a membrane laid loose over the roof deck and not fixed down to it. This membrane is ballasted with shingle or soil and turf, protecting it from the ultra-violet radiation from the sun and from mechanical damage. It keeps the membrane at a steady temperature, and avoids puddles of water forming (which cause differential temperatures at the edge of the puddle, eventually leading to the membrane failing). The edge is retained by a capping, whilst still allowing the membrane to move independently of the deck to accommodate movement arising from changes in temperature and the moisture content of the deck. Adequate ventilation is needed. A flat-roofed building can be extended easily in any direction.

Roof structure

Timber is generally used for pitched-roof construction, and is also the preferred material for flat-roof construction. Other options include prefabricated roof cassettes and composite I-beams. Care has to be taken that the panels used have minimum environmental impact: softwood ply, OSB or low-formaldehyde chipboard are all acceptable; ply from tropical hardwood is not. Trussed rafters at 600mm centres use less timber than cut rafters on purlins, and this is the most economical form of roof construction. However, it has two disadvantages. Firstly, the loft formed by the trussed rafters has to be well ventilated to avoid condensation; this introduces cold air into the building, which then has to be kept out of the house at ceiling level – which can be difficult as there will be many penetrations at ceiling level for light fittings, water pipes, vent pipes, ventilation ducts,

A mezzanine for storage or sleeping gallery replaces a conventional cold and draughty loft.

ROOF STRUCTURE				
First preference	**Second preference**	**Third preference**	**Not recommended**	**Comments**
Softwood or I-beam rafters Prefabricated timber panels	Trussed rafters Corrugated steel Pre-cast concrete planks	Concrete with recycled aggregate	Solid concrete	

flues and loft access, all of which must be sealed. The other disadvantage is that the roof space is unusable for anything other than storage. The preferred arrangement is for the insulated and airtight envelope to follow the pitch of the roof. This creates interesting spaces under the roof, and mezzanine sleeping or study galleries can be introduced.

Flat roofs can also be constructed using a lightweight corrugated steel deck. This is very thin and therefore uses little material, but does require galvanizing with zinc to protect it against corrosion. Concrete is also used, but should contain a proportion of recycled aggregate. Pre-cast concrete planks with hollow cores are a more efficient use of material.

Roof insulation

Insulation in the roof has to do two things, as elsewhere in the building: it has to keep the inside of the building warm whilst also allowing moisture vapour which is generated inside the house from cooking, bathing and breathing to escape to the outside without causing damage through condensation.

This can be achieved in two basic ways:

- A 'cold' roof, which relies on ventilation to prevent the build-up of moisture, or

- A 'warm' roof, which relies on an impermeable vapour control layer preventing condensation.

In a cold roof, the ventilation is on the cold side of the insulation. This is commonly in the form of a ventilated loft space in a pitched roof. If the ceiling follows the roofline either in a flat or pitched roof, then a ventilation gap has to be provided between the insulation and the roof covering. The advantage of this arrangement is that any type of insulation

ROOF INSULATION				
First preference	**Second preference**	**Third preference**	**Not recommended**	**Comments**
Cork Sheep's wool Cellulose fibre	Expanded polystyrene Mineral wool Foamed glass		Polyurethane Extruded polystyrene	Cork for warm roof Mineral wool can be quilt for cold roof but has to be rigid slabs for warm roof Extruded polystyrene for upside-down roof

PITCHED-ROOF COVERINGS				
First preference	**Second preference**	**Third preference**	**Not recommended**	**Comments**
Thatch Timber shingles Green roof	Clay or concrete tiles Natural UK slate Timber boarding	Artificial fibre, cement slates and corrugated sheets Bituminous shingles Coated steel Stainless steel Aluminium	Copper Zinc	Recycled slates or tiles are preferred if they are available

material can be used, including insulation from renewable sources. It also permits a 'breathing' construction which is free of the risk of condensation. It is vital that the ventilation is adequate, but the introduction of ventilation does make it more difficult to ensure that the building is airtight against heat loss.

In a warm roof, the insulation is on the outside of the structure which is therefore protected from any risk of dampness or condensation. An effective vapour-control layer is required on the warm side of the insulation which can be difficult to achieve. The insulation has to be in the form of rigid boards, which tends to rule out many renewable materials such as wool and cellulose fibre, although cork is a renewable if somewhat expensive alternative.

A variant of the flat warm roof is the 'upside-down' roof which uses waterproof insulation above the waterproof membrane. This is then ballasted down with paving slabs, shingle or perhaps turf to prevent the lightweight insulation being sucked off the roof by uplift from the wind. This avoids any risk of condensation on the underside of the waterproof membrane, which is now on the warm side of the insulation. However, it does require extruded polystyrene as the insulation material which is an environmentally undesirable material.

Pitched-roof coverings

Timber shingles and thatch are both renewable coverings for pitched roofs, and are preferred from an environmental point of view. Thatch has reasonably good insulation properties, and is

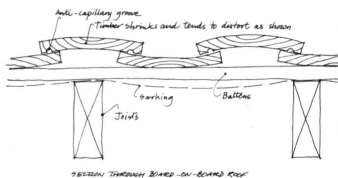

Anti-capillary groove
Timber shrinks and tends to distort as shown
Sarking   Battens
Joists

SECTION THROUGH BOARD-ON-BOARD ROOF

Thatch is a renewable roofing material with moderate insulating properties.

around 400mm thick and may provide the necessary thermal performance without any additional insulation material. Reed thatch lasts longer than straw, but costs more. Local oak split shingles are preferred to imported cedar sawn ones, which may be more available. However, sources of cedar are under threat, so make sure that cedar shingles are from a certified sustainable source. Board-on-board durable timber, often natural larch boards, can also be used as an economical roof covering – this is common in Scandinavia. Boards should be laid so that they tend to cup alternately upwards and downwards as they dry out, to prevent gaps opening up, and a good breathable membrane underlay is required to prevent leaks. A groove in the underside of the edge of the top boards reduces the risk of rain penetration. Consideration will have to be given to the combustibility of these coverings, which may

have to be separated from other buildings.

Clay or concrete tiles are relatively good and readily available. Concrete is cheaper than clay. Natural slate is also good, but much slate is now imported from Spain and Belgium, which increases its environmental impact.

Aluminium, on the other hand, is a very high-energy material; coated steel is galvanized with zinc; and stainless steel contains heavy metals chromium and nickel; all of these have significant environmental impacts. Fibre-cement slates and corrugated sheets contain more cement than tiles; resin-bonded slates are energy-intensive in manufacture, and bituminous shingles are a product of the petrochemical industry.

Both zinc and copper cause damage to the environment by corrosion contaminating water runoff and thus the soil. There are alternatives to both, and they are not recommended.

Green roofs can be used up to a pitch of around 30^0 providing steps are taken to prevent the soil sliding down the roof. This can be in the form of netting or a framework of battens incorporated into the soil.

Flat-roof coverings

There is no renewable flat-roofing material. EPDM (see page 268), which is artificial rubber, is the least environmentally damaging synthetic membrane. Some EPDM membranes contain a proportion of recycled rubber. Also good is Thermoplastic Polyolefin (TPO) also described as Flexible Polypropylene/polyolefin Alloy (FPA). Modified or

FLAT-ROOF COVERINGS				
First preference	Second preference	Third preference	Not recommended	Comments
	EPDM		PVC	
	TPO		Zinc	
	Modified bitumen felt		Lead	

FACING PAGE Left: This shed has a board-on-board roof of Douglas fir, and is an economical roof covering using renewable materials. Right: Timber boards tend to 'cup' away from the heart of the tree as they dry and shrink, and they should be fixed alternately as shown to avoid rain penetration.

FIXING FLAT-ROOF COVERINGS				
First preference	**Second preference**	**Third preference**	**Not recommended**	**Comments**
Loose with plants	Loose with gravel	Mechanically fixed	Fully bonded to deck	Allow for additional weight of ballasted roofs

high-performance felt is formulated to resist ultra-violet radiation, and is longer-lasting than basic bituminous felt. It is thinner, and often only requires a single layer rather than a three-layer build-up required with basic bituminous felt. The pollution caused by the production and waste-processing of PVC is substantial. The production of both zinc and lead is relatively harmful, the metals are relatively scarce, and they cause pollution of ground water when used externally.

Fixing flat-roof coverings

Loose-laid roof membranes perform better than fully bonded ones, and increase the opportunities for recycling. Ballasting the roof protects the membrane from physical damage and the effects of ultra-violet radiation, which lengthens its lifespan. Ballasting with sedum on clay granules is better than gravel, which is in relatively short supply. A green roof buffers storm water from surcharging the drainage system, modifies the microclimate in towns, and also creates habitats for wildlife. However, you will have to increase the load-carrying capacity of the roof structure to carry the additional weight. Mechanically fixed systems are better from a recycling point of view than systems which rely on bonding the membrane down to the deck with adhesive.

Flashings

Lead and zinc are commonly used for roof flashings, but are both harmful to the environment (see above). Zinc is in short supply, and does not have a very long life. Synthetic alternatives exist, but tend to be more expensive. They are also light in weight, and therefore liable to lifting by the wind. Adhesive flashing materials ('Flashband' for example, which is made from aluminium with a bitumen backing) can overcome this problem but could lead to difficulties with separation for recycling at disposal stage. Stainless steel of a quality that is soft enough to work and coated to look like lead may be an alternative, but it is expensive; or flexible aluminium, although this is a very energy-intensive material. The use of lead may be unavoidable in certain circumstances.

FLASHINGS				
First preference	**Second preference**	**Third preference**	**Not recommended**	**Comments**
Polythene EPDM		Coated stainless steel Aluminium	Lead Zinc	

WINDOWS AND EXTERNAL DOORS				
First preference	**Second preference**	**Third preference**	**Not recommended**	**Comments**
Sustainable durable wood	Softwood with boron-based implants	Aluminium Preserved softwood Steel Aluminium / timber composites	PVC Uncertified hardwood	

Windows and external doors

PVC is the most environmentally damaging option, followed by aluminium. Softwood was widely used in the 1960s and 70s, but the frames were often poorly designed and poor-quality wood was used; and so softwood windows have a poor reputation. Recent research by The Building Research Establishment suggests that PVC windows, which are often thought to have low maintenance costs are, on the contrary, easily damaged and difficult to repair, and have a relatively short life. They are not a good alternative to good quality timber windows. Modern designs are based on Scandinavian practice and avoid the principal problems. They do not have projecting timber sills, and use aluminium for the vulnerable bottom bead holding in the glass. The sections are more substantial and are manufactured of redwood. However, they are generally impregnated with preservative. Certified sustainable hardwood is durable: locally sourced oak or sweet chestnut is best, although imported tropical hardwood is good. The vulnerable parts of an otherwise untreated softwood frame, the bottom corner joints, can be protected by solid boron implants. This benign treatment dissolves into the surrounding wood if and when it gets damp and there is a risk of decay. Sustainable plywood is a good solution for external doors.

Steel windows made of either hot-rolled or cold-rolled sections have to be zinc-galvanized, and both steel and aluminium windows have to have a thermal break in the sections to prevent cold bridging and condensation problems, which puts up the cost.

Timber windows can be fitted with clip-on aluminium extrusions on the external face to provide a

WINDOW SILLS				
First preference	**Second preference**	**Third preference**	**Not recommended**	**Comments**
Brick Natural stone	Pre-cast concrete Artificial stone	Aluminium Fibre-cement boards		Aluminium or fibre-cement boards for windows in lightweight cladding systems

INTERNAL DOORS				
First preference	**Second preference**	**Third preference**	**Not recommended**	**Comments**
Hardboard-faced hollow core	Softwood	Sustainable plywood	Tropical hardwood Steel Aluminium	

very low-maintenance finish. However, the energy intensity of the aluminium means that this type of window does badly as far as environmental impact is concerned. The maintenance of timber is less critical if sustainable hardwood or softwood – together with a factory-applied microporous paint finish, which can be expected to last 15 years before it needs to be redecorated – are specified.

Glazing

Sealed double-glazing with low-emissivity coating, argon-filled cavity and with thermally broken spacers is preferred. The energy savings outweigh the environmental effects of the gas filling and coating. A low-emissivity coating is a microscopic thickness layer on the glass which reflects infra-red radiation back into the room. They are available in two types: so-called 'hard' coatings and the more efficient 'soft' coating which is preferred. Argon is an inert gas which is much more viscous (syrupy) than air, which cuts down the rate at which heat is transferred across the cavity in double-glazing. Spacer bars form the edge to a sealed double-glazing unit, and are generally made of aluminium. Once you have an efficient double-glazing system with low-emissivity coating and an argon-filled cavity, the heat loss through the cold bridge formed at the edge by the aluminium spacer bar becomes significant, and can cause condensation to form, which will cause the finish to the window to deteriorate and could even cause rot in a timber window frame. The solution is

to use a spacer bar with a thermal break of plastic between an inner and outer aluminium section.

Dry glazing tape is preferred to traditional wet putty or sealant, which has a shorter life and requires more maintenance. Dry glazing makes recycling glass easier.

Window sills

If timber sills are not incorporated into the window, separate sills will have to be provided. Brick and natural stone are good from an environmental point of view; pre-cast concrete and artificial stone less so. Aluminium and fibre-cement sheet are still less desirable, but are necessary choices for windows in lightweight cladding systems.

Internal doors

A hollow-core door made with a honeycomb of cardboard faced with hardboard rather than plywood is the preferred product from an environmental point of view. A softwood solid timber door requires more material. Plywood, including ply which has been veneered with softwood to give a 'pine' finish or similar, is more often than not manufactured from tropical hardwood. A melamine layer is also often applied. The lipping on flush doors is also more often than not tropical hardwood. Thresholds are best in sustainable hardwood such as oak for example.

STAIRS				
First preference	**Second preference**	**Third preference**	**Not recommended**	**Comments**
Timber	Steel	Concrete with recycled aggregate	Tropical hardwood	Durable species of timber are required externally

Stairs

Softwood is the preferred choice. Care needs to be taken because standard staircases generally have ply risers, which are often made of tropical hardwood. Also the strings are often made of Pirana Pine which is an endangered tropical timber species.

In certain circumstances a non-combustible stair may be required. Steel is damaging to the environment because of the extraction of coal and iron ore and the pollution generated during manufacture. Concrete aggregates are in short supply and their extraction damages the environment, and the production of cement uses large quantities of fossil fuels and produces much pollution. A proportion of recycled aggregates should be used if possible. Tropical hardwood should not be used unless it is certified as from a sustainable source.

Durable, sustainable timber, such as oak, is the preferred solution for external stairs. Treated softwood should not be used because of the harmful effects of timber preservatives on the environment.

Steel is often used for external fire escape stairs, but it requires protection against corrosion.

Sealing joints

The need for sealing joints can often be reduced by design. Seals made from renewable raw materials such as coconut fibre, felt or sisal are not readily available and are not suitable for joints exposed to the weather. Polyethylene foam tape, mineral wool and EPDM synthetic rubber seals have relatively low environmental impact with regard to their extraction and manufacture, and are the suggested alternatives. Depending on the requirement for the seal, you could also use an elastomeric sealant on a backing of mineral wool. Polyurethane foam or PVC foam tape are not recommended. Elastomeric sealants such as silicon or polysulphide sealants last longer than plastic sealants such as acrylic sealants.

The selection is based on the situation, the width of the joint, whether it has to withstand moisture and the amount of movement that is anticipated.

SEALING JOINTS – BACKING				
First preference	**Second preference**	**Third preference**	**Not recommended**	**Comments**
Renewable materials – coconut fibre, felt or sisal – not readily available or suitable for joints exposed to the weather	Polyethylene tape Mineral wool EPDM seals	Elastomeric sealant on backing	Polyurethane foam PVC foam tape	Selection based on situation

SEALING JOINTS – SEALANT				
First preference	**Second preference**	**Third preference**	**Not recommended**	**Comments**
Silicon sealant	Polysulphide sealant Water-based acrylic sealant		Polyurethane sealant Solvent-based acrylic sealant	Selection based on situation

Paintwork

Natural surfaces that do not require an applied finish are preferred as they minimize maintenance and the lifetime environmental impact. Thus brick,

stone, durable timber and 'through colour' lime render can all be left without an applied finish.

Natural paints and stains are made from renewable resources, and are preferred to common paints which are derived from petrochemicals. However, most natural paints do still contain solvents, which can be harmful, particularly to the painter, when working indoors. Water-based paints, on the other hand, have a much lower solvent content than other paints, and for this reason are preferred for indoor use. High-solid versions of common paint contain less solvents. Natural paints, stains and varnishes tend to be more expensive. Mineral paint for external masonry binds with the background chemically and is very long-lasting.

External steelwork

External steelwork has to be protected from corrosion. A coating of zinc is often applied by dipping in a bath of molten metal. This zinc galvanizing leaches when water flows over it, and in

PAINTWORK				
First preference	**Second preference**	**Third preference**	**Not recommended**	**Comments**
Natural paint, stain or varnish Mineral paint for external masonry or render	Water-based paint	High solids alkyd paint	Alkyd paint	

A wide range of natural paints, stains and clear finishes is now available.

EXTERNAL STEELWORK FINISHING				
First preference	**Second preference**	**Third preference**	**Not recommended**	**Comments**
Powder coating	High-solids alkyd paint	Alkyd paint Red lead	2-part epoxy paint systems Galvanizing	

this way harmful zinc finds its way into watercourses and the soil. The zinc layer can be damaged, particularly when tightening fixings. The galvanizing can be painted over, but this requires an environmentally damaging primer. This disadvantage can be avoided by applying a powder coating over the galvanizing; this is applied electrostatically in the factory, which avoids the use of solvents. The powder coating is very durable, and does not require maintenance during its lifetime, although it can be difficult to repair a damaged coating. High-solids alkyd paint contains less solvents than common alkyd paint. Red lead is now manufactured without being based on the use of lead because lead is so damaging to human health. 2-part epoxy paint systems involve harmful emissions during manufacture.

Kitchen fittings

Sustainable timber is the preferred material for kitchen units. Plywood made of sustainable timber lasts longer and contains less resin than chipboard.

KITCHEN FITTINGS				
First preference	**Second preference**	**Third preference**	**Not recommended**	**Comments**
Laminated beech	Stainless steel Natural stone	Synthetic resin board	Melamine-faced chipboard	Worktops
Sustainable timber	Sustainable plywood	Zero-formaldehyde MDF	Uncertified tropical plywood	Cabinets

Kitchen units made of zero-formaldehyde Douglas fir plywood.

FLOOR FINISHES				
First preference	Second preference	Third preference	Not recommended	Comments
Linoleum Natural cork Natural stone	Ceramic tiles		PVC flooring PVC-faced cork tiles	Mortar bedding is preferred for ceramic tiles Use natural adhesives

Zero-formaldehyde chipboard is now available. Chipboard is generally surfaced with a melamine film, which complicates disposal. Plywood made from uncertified tropical hardwood should be avoided. Timber kitchen units are very much more expensive than chipboard.

Laminated beech work surfaces are preferred, as they are manufactured from renewable resources. Stainless steel can be easily separated from its backing and reused at the end of its useful life. Natural stone is not renewable, but it is a plentiful resource. Some materials such as marble and granite, however, may be imported from afar. Synthetic resin board materials such as 'Corian' cause problems with waste disposal. Melamine-faced chipboard is the most common worktop material by far. but is the most environmentally damaging.

Floor finishes

Linoleum is manufactured from renewable resources: linseed oil, cork and jute. Stone is good, and should be locally sourced if possible. Natural cork is good, but beware: most cork floor tiles sold in Britain have a PVC coating, which is not good.

Unfinished natural cork tiles are surprisingly hardwearing if treated with two coats of sealer.

Ceramic tiles require a large quantity of energy to manufacture. A mortar bedding is preferred for fixing ceramic tiles on a solid floor. Natural adhesives are available based on renewable resources, and water-based adhesives are also available with much reduced amounts of solvents used. Solvent-based adhesives are to be avoided, as should PVC flooring. Linoleum costs more than PVC, but lasts longer.

Drainage

The preferred material for below-ground drainage is vitrified clay, followed by concrete (in spite of its cement content and the pollution that this implies). Polypropylene is often used for the flexible connections and inspection chambers of clay drainage systems, and this is much less environmentally damaging than PVC, which should be avoided.

The preferred material for above-ground drainage is polyethylene, which is now available in the UK. PVC, which is the generally available product, is to be avoided.

DRAINAGE

First preference	Second preference	Third preference	Not recommended	Comments
Vitrified clay for below ground drainage	Polyethylene for above-ground drainage and waste systems	Concrete for below-ground drainage	PVC	

RAINWATER DRAINAGE

First preference	Second preference	Third preference	Not recommended	Comments
EPDM-lined timber gutter	Coated and galvanized steel	Coated aluminium	PVC Zinc Copper	It may be possible to design the roof so that no gutter is needed

Rainwater drainage

A timber gutter lined with a EPDM lining is the solution with the least environmental impact, although it is more expensive than other alternatives. Lead, zinc and PVC linings are to be avoided. Another alternative is to design out the need for a gutter by using a green roof with a large overhang with a strip of shingle below which can either lead to a drain or soakaway. If you need an applied gutter, the least environmentally damaging material to use is coated and galvanized steel followed by coated aluminium. PVC, zinc and copper are to be avoided.

PIPEWORK

First preference	Second preference	Third preference	Not recommended	Comments
Polyethylene Polybutylene Polypropylene	Stainless steel	Copper		Polyethylene for external mains Stainless steel for exposed internal pipework

Above: Galvanized steel rainwater goods are common in Europe, and much preferred to PVC.
FACING PAGE Left: Natural cork floor tiles with an applied sealer finish, but note that most tiles are sold with a PVC coating.
Right: Polyethylene is preferred to PVC for above-ground drainage.

HEATING AND HOT WATER				
First preference	Second preference	Third preference	Not recommended	Comments
Condensing boiler with solar heater for hot water	Condensing boiler		Standard boiler Electrical heating	

Services

Pipework

Synthetic water pipes do not corrode, and their manufacturing processes are less damaging than the alternatives of stainless steel or copper. Copper is subject to corrosion, which leads to contamination of waste-water with the metal, which is damaging to water organisms in particular. Polyethylene can only be used for cold-water pipes, and blue polyethylene is now general for external water mains. Plastic pipes are now used widely in push-fit plumbing systems, which make plumbing a much less skilled job: push-fit systems are quick and foolproof. Stainless steel can be used in push-fit systems, and this is a good choice where pipework is exposed to view.

Heating and hot water

Condensing gas boilers are preferred because they are more efficient, use less non-renewable energy and pollute less than conventional boilers (which are now not permitted under the Building Regulations). A solar hot-water system is also preferred, as it also reduces fossil-fuel use. The use of electricity for heating and hot water is not recommended, because electrical energy is not efficient and creates a large amount of pollution at the power station. Gas boilers should be designed to minimize NOx emissions, and if a combination boiler is used (one that delivers instantaneous domestic hot water as well as supplying hot water for the heating system – a much simpler installation, as no hot-water storage tank is required and it can be more efficient especially for smaller dwellings) it should not have a permanent 'keep warm' facility which is designed to keep a small amount of hot water ready for use at all times, as this reduces overall energy efficiency.

Electrics

Cables with low smoke zero-halogen (LSOH) insulation should be preferred to PVC-insulated cable, which is more environmentally damaging in its production and produces lethal fumes in a fire.

Ventilation

The most energy-efficient method of delivering the

ELECTRICS				
First preference	Second preference	Third preference	Not recommended	Comments
LSOH cable			PVC-insulated cable	

VENTILATION				
First preference	**Second preference**	**Third preference**	**Not recommended**	**Comments**
MVHR	PSV	MV with trickle vents	Air conditioning	

fresh air necessary to remove odour and indoor pollutants is Mechanical Ventilation with Heat Recovery (MVHR). This extracts stale air from the rooms in which it tends to be generated – the kitchen and bathroom – and supplies fresh air to the habitable rooms – the living-room and bedrooms. This fresh air is pre-heated in a heat exchanger, using the warm stale air that is being extracted to minimize the heat loss associated with ventilation. The supply of fresh air can be filtered to remove pollution and dust, but the filters have to be cleaned at least once a year. If the heat load is small, it may be possible to fit a heater in the supply duct and use this as the heating as well as the ventilation system. This system implies a substantial amount of ductwork which has to be incorporated into the building. It is important to ensure that the power used by the fan is low (low wattage DC motors are often used rated at 6w) and certainly no more than the heat recovered from the stale air.

A whole house mechanical extract or positive input system is similar but only has one set of ducts, either on the supply or extract side, and is supplying or drawing in cold fresh air which then has to be heated, thus using more energy. It is simpler and cheaper, however. All these mechanical systems require a certain amount of maintenance of the fan.

Passive Stack Ventilation (PSV) works on the principle that warm stale air rises. Ducts rise from the kitchen and bathroom, drawing in fresh air through vents in the habitable rooms. The rate of extract and input of fresh air are controlled by humidity-sensitive extract and inlet grilles: more air is extracted when you cook or take a bath, and more fresh air is sup-

plied to rooms which are occupied. The system works without any energy input, and requires little or no maintenance as there are no moving parts. It can be difficult sometimes to arrange ducts with the necessary rise – in a flat for instance – and in certain wind conditions the system does not work properly.

The simplest system is to extract stale air from kitchens and bathrooms with a local fan in each, and to provide so-called 'trickle vents' in the habitable rooms. These are often in the form of a slot in the head of the windows. It is a very simple and economical system, but it is not very controllable. Humidity-sensitive controls can be fitted so that air is only extracted when the humidity level rises, and individual heat-recovery fans can be used which recover heat from air extracted from the kitchen and bathroom locally.

A low-tech alternative to a heat-recovery system is to draw fresh air from a conservatory where it will be pre-heated by solar radiation at certain times, and thus reduce the heating demand. Similarly, there are systems on the market which draw fresh air from the loft, which will be warm at certain times, or through a box below the roof covering intended to capture heat, although it is not clear how effective this strategy is in practice.

With growing concern over global warming, there is a possibility that people in the UK will start to expect the ability to reduce temperatures on hot days in summer to more comfortable levels. People have become accustomed to air conditioning on holidays in hot climates, but using electricity for cooling is as environmentally damaging as using it for

LANDSCAPING				
First preference	**Second preference**	**Third preference**	**Not recommended**	**Comments**
Hedges	Sustainable durable timber	Masonry	Uncertified tropical timber Treated softwood	

heating. Comfort can be improved by shading windows, providing good insulation, incorporating high thermal mass, and through ventilation. Air movement can be induced by exploiting the stack effect, and passive methods can be used to cool incoming air by using earth tubes (pipes buried in the garden which cool incoming air because the soil is at a lower temperature than the air).

Landscaping

The least environmental impact solution to garden enclosures is to plant hedges rather than build fencing. Hedges contribute to the microclimate and biodiversity by providing natural habitats. A number of native shrubs are suitable, including hornbeam, hawthorn and blackthorn. A temporary fence may be required until the hedge becomes established. Sustainable durable timber is an alternative which

has less impact that masonry walls. Fixing the posts on concrete spur posts avoids decay where timber is in contact with the ground, and avoids the need for timber treatment. Non-certified tropical hardwood and treated softwood should be avoided.

Paving

Reducing the area of hard paving as much as possible conserves resources and reduces demand on the capacity of the drainage system, reduces the risk of flooding, and returns water to the soil. Permeable, natural materials such as gravel or wood chips are preferred. Concrete slabs are preferred to clay paviors; concrete slabs containing a proportion of recycled aggregate are better still. Asphalt is to be avoided, as it involves the extraction of aggregates and the emission of sulphur dioxide, nitrogen dioxide, volatile organic compounds during production, and fumes during laying.

PAVING				
First preference	**Second preference**	**Third preference**	**Not recommended**	**Comments**
Concrete slabs with recycled aggregate Gravel or hoggin Wood chips or bark mulch	Concrete slabs	Clay paviors	Asphalt	

Designing a sustainable garden

A sustainable home should be set within a sustainable garden, with high ecological value which supports wildlife.

The need for biodiversity

There is a real need for retaining existing and creating new habitats in Britain. The statistics on the loss of UK habitats since 1949 speak for themselves: lowland meadows 95%, lowland heath 50-60%, ancient lowland woodlands 30-50%, lowland wetlands (fens) 50%, hedgerows 25%, upland heaths and grasslands 30%. The impact on wildlife has been severe: for example, the populations of some UK woodland birds, such as the song thrush, have fallen by more than half since the mid-1970s. The loss of the house sparrow is a sad urban case study: the species has declined by 95% since 1990.

Current planning policies preventing development in the green belts around our cities and limiting development in other rural areas to infilling existing village areas are intended to slow down or stop the destruction of natural habitats and the wildlife they support, as well as preventing loss of land used for agriculture. However, urban brownfield land or derelict sites which have been left to regenerate naturally with little or no human interference are often ecologically richer than intensively managed agricultural land which is treated with pesticides and fertilizers. Brownfield land may be occupied by often-hidden, rare, protected or locally important species. Care should be taken when

developing brownfield sites, as well as preserving the natural habitat of greenfield areas.

It is important to consider the relationships between wildlife (birds, mammals, insects, butterflies, amphibians) and their ideal habitats (plants, hedges, trees, grasslands, ponds and ditches), and to create habitats for wildlife in both the country and the town.

A sustainable garden

A sustainable garden:

- protects existing habitats and provides new natural habitats

- sustains biodiversity with wildlife alongside human occupants

- connects into and integrates with the wider ecology of the area

- is productive and provides opportunities for recreation and sensory delight.

Soils and materials

A sustainable landscape will maximize the use of on-site resources. You could commission an ecological survey to determine what ecological resources exist already, and to provide advice on how to preserve and enhance them. This could range from avoiding the importation of topsoil to having a comprehensive water strategy for the site.

- Avoid or limit the importation of topsoils, by creating topsoil from site-won subsoil and crushed rubble combined with organic composts or waste materials such as mushroom compost and horse manure. The finished soil profile may not initially be as well structured as an organic imported topsoil, but can support and establish native plantings. If managed creatively, constructed topsoils

will develop a natural soil structure relatively quickly.

- Minimize hard landscaping surfaces as far as possible; maximize soft and permeable surfaces.

- Retain and use water on site.

Landscaping materials (e.g. bricks and paving, and timber) should be selected using the same criteria as building materials, so use materials of low embodied energy, locally sourced, with a preference for reclaimed or recycled materials.

Plant species

A suburban garden can be sterile and static, merely forming a decorative backdrop to the home. Such cosmetic planting schemes use ornamental species with a lifespan of 10-15 years. There are considerable environmental benefits in providing more sustainable forms of planting, combining native and carefully selected non-native/exotic plants:

- Select species appropriate to microclimate, soil types and locally occurring woodland types. From these, choose plants that are most suited to native wildlife and insects.

- Use native species; seek local guidance from experts (native species are cheaper in establishment and maintenance and support more wildlife species).

- Use plant stock from the locality, as it is better suited to its habitat.

- Plant oak, willow, silver birch, hawthorn, alder, aspen, and rowan as these trees support the most insects. Note that sycamore, Norway maple, horse chestnut, sweet chestnut, and Japanese cherry are not native species and support very few insects.

FACING PAGE *Native wild flowers support insects and other wildlife and help maintain biodiversity.*

Creating a sustainable garden will provide diverse wildlife habitats and encourage biodiversity. A sustainable garden will, on the whole, be less expensive to establish and maintain than a more exotic landscape.

- Consider getting and growing plant stock in advance so that you can move into an established garden.

- Plant new trees and have a tree-planting programme.

- Create a pond and plant it with species that thrive in a wet environment.

- Consider green roofs: they can replace greenery lost at ground level if they are planted as meadow, and do not need watering. Also, extensive roof gardens can be based on non-native sedum species or herb species, both of which can provide a rich food source for insects.

Maintenance

We need a change in culture regarding the type of landscape we expect around our homes. People often expect neatly manicured lawns and weed-free flowerbeds with neat rows of annuals or large areas of chemically maintained grassland using high-toxicity pesticides and artificial fertilizers. Such forms of landscape give low ecological value and are expensive to maintain.

Costs and benefits

Creating a sustainable garden will provide diverse wildlife habitats and encourage biodiversity. A sustainable garden will, on the whole, be less expensive to establish and maintain than a more exotic landscape. So whilst obtaining an ecological survey adds to costs, it is relatively inexpensive. Also, maximizing site-won resources and planting native species give cost savings. Habitats requiring low or poor fertility soils can avoid or limit the cost of importing topsoil. Sustainable landscapes are often less labour-intensive to construct, and maintenance is cheaper as it is less intensive, does not use chemicals, and keeps watering costs low.

- Preserve existing mature trees: they take a long time to grow, provide instant visual effect and foster wildlife. Trees within Areas of Outstanding Natural Beauty and Conservation Areas will have statutory protection. Provide protection to established trees and plants during the construction work.

- Consider relocating existing vegetation that is at risk. For example, large sections of existing hedgerow can be and have been successfully transplanted.

Summary

When developing a new garden for your new home:

- Contact the local wildlife trust which is often a good first point of contact for information about wildlife in the area.

- Adopt a wildlife plan and try to provide foliage to compensate for that lost in construction where existing green areas have been built on.

- Develop the site in a way that protects and enhances the most important ecological attributes.

- Provide habitats that will protect and foster local wildlife.

- Develop a planting strategy that combines native species with carefully selected non-natives.

- Maximize the use of locally occurring or site-based resources such as soils.

- Retain water on site.

- Create a pond.

- Choose landscaping materials for paths and walls on the basis of low embodied energy.

- Consider long-term maintenance implications: think about minimal mowing regimes, avoid sprays and pesticides, encourage natural pest control through companion planting, encourage minimal use of water and fertilizer, specify organic alternatives to peat and organic forms of fertilizers and pesticides – this can often reduce maintenance costs.

Why we should build green for the future

The prospect of people planning and building their own adaptable, low-environmental-impact houses is a good model for the new homes we need for the 21st century. More and more people are building for themselves, which is good for them and good for society as a whole — but let us look forward to the day when most of those people are building energy-efficient, 'green' homes.

The prospect of people planning and building their own adaptable, low-environmental-impact houses is a good model for the new homes we need for the 21st century.

Current housing is not sustainable. Britain has a unique reliance on mass housing, which is generally not able to deliver homes of high quality, adaptable to changing needs and expectations and with a long useful life – in other words, which is not sustainable. The mass market tends to serve the interests of developers rather than individual occupants or the common good. It does not offer opportunities for residents to be part of a sustainable housing system, it does not value adaptability or innovation, nor higher standards of performance or construction. The products of the volume house-building companies do not offer good value for money.

Meanwhile, although Britain is committed under the Kyoto Protocol to cutting our greenhouse gas emissions and the level of emissions has gone down, it is now rising again. Transport sector emissions continue to grow, with more and more vehicles on the roads, cheaper air flights, North Sea gas running out faster than predicted – and so more electricity is generated using coal which is a more polluting fuel than gas. Current measures are only scratching the surface of the issue. The government is committed to a 20% reduction in CO_2 emissions by 2010 under Kyoto arrangements – which it is now clear will not be achieved – and to the Royal Commission on Environmental Pollution target of a 60% reduction by 2050, which the Intergovernmental Panel on Climate Change estimate can be achieved using existing technologies but which does, nevertheless, imply changes in the way we use energy. Some experts are of the view that a reduction of 80% is required by 2050 in order that concentrations of CO_2 in the atmosphere stay within safe limits. Think about it: this means reducing the carbon footprint of every citizen in the UK to one-fifth of its current level – and it is more difficult to reduce carbon emissions in some areas such as electricity gen-

eration and transport than in others – so emissions from the domestic sector and buildings in particular may have to be reduced to a still greater extent to compensate.

Therefore, building in a more sustainable way is not an issue that is going to go away – momentum is in fact gathering, pushed by international standards. For example the European Union has introduced a number of significant Directives over the last year or two, aimed at improving waste management, resource use and the use of energy in buildings in particular – all of which have fundamental implications for the design and construction of buildings and homes. These Directives are incorporated into recent national strategies which have addressed the creation of sustainable communities, how to promote sustainable construction and the role of renewable energy. The new Regional Development Agencies are charged with establishing regional plans for different aspects of the economy, including housing and the environment, and they have regional strategies to promote sustainable development. Many local authorities have local policies for sustainable development, arising often from Local Agenda 21 initiatives and also from the desire on the part of local councils to promote a higher standard of development in their area.

Public opinion is shifting towards wanting energy efficiency and 'green' values. Whilst price and location remain the principal reasons behind housing choices, there is a significant niche in the market for 'green' homes, and a few developers are responding to this demand. The mass of the house-building industry is slow to change, and does not offer the choices many people want. In this context, the government has proved to be reluctant to legislate for change – rather preferring to support the conservative approach of the house-builders by setting standards at the lowest level consistent with minimum international compliance.

Tougher legislation is only useful if people com-

Self-builders together now build more houses than the largest individual house-builder.

developers have successfully built and marketed homes with ecological features at a premium, and this has encouraged larger house-builders to make moves in this direction. There are also examples in the public sector of low-environmental-impact housing championed by a handful of forward-looking organizations.

Individual householders, generally speaking, want more than industry or standard government policy provides. People have become more aware of standards elsewhere on the Continent and in North America, and their expectations are rising. They want choices over standards of space and equipment, and they want higher standards of energy consumption and environmental performance than a mass-produced house offers. More and more people are commissioning or building their own home. This sector now accounts for around 25,000 dwellings per annum – about 15% of the new homes built every year – which has grown from around 4,000 per annum in 1980. Self-builders together now build more houses than the largest individual house-builder.

However, Britain still has a uniquely small self-help sector in the housing market – it is twice as large in the USA, Australia, Japan and other developed economies, and four times as large in Germany, where houses built speculatively for sale are not that common. In less developed economies, building your own accommodation is often the only way to obtain a place to live. Indeed it is only relatively recently, with the rise of the industrial city in Britain, that people have not been in control of their own housing. Philanthropists developed model industrial dwellings as a response to the excesses of the Victorian city, and when this did not meet the demand for decent homes, the local authorities started building. There was a boom in private speculative building between the wars, and mass-housing provision by local authorities and then housing associations and private house-builders has been the norm in the post-war era.

ply with its requirements – either voluntarily or as a consequence of enforcement action – and this is where the government is unwilling to act, either for fear of additional expenditure or because it does not want to antagonize industry or the public. Developers are willing to comply, provided there is a 'level playing field' which requires all their competitors to comply equally. A recent survey by the BRE, however, showed amazingly that only two out of three new houses complied with the relatively low minimum standards of the then current Building Regulations. The building control system is overstretched in many areas, with insufficient staff often with inadequate training. Tougher standards without stronger enforcement undermines the private market, and the government has therefore not supported this approach of stronger legislation to force improved standards of energy and environmental performance. Rather, they have used legislation to bring along the laggards and raise standards at the bottom of the market. This approach relies on expectations rising in the private market to lead the way towards higher standards; a significant number of individual eco-houses have been built, and a handful of small

A more significant self-help sector would tend to reduce the very high price of housing in Britain by undercutting the cost of mass-produced housing. It would create choice for residents to better match their home to their wants and needs.

Advantages of a significant self-help sector

A more significant self-help sector would create a number of benefits for the housing market. It would tend to reduce the very high price of housing in Britain by undercutting the cost of mass-produced housing. It would create choice for residents to better match their home to their wants and needs. The quality of homes would tend to rise, driven by a desire for higher standards of performance and equipment, and this would also tend to improve the energy performance and the general sustainability of homes. Encouraging the self-help sector would encourage innovation and energy efficiency.

Participation by residents is a necessary feature of any sustainable housing system. It is important that residents support the principles and understand how to operate their sustainable home efficiently. It has also become more recognized by the government and others that self-build could have a role in the process of regenerating run-down neighbourhoods and estates. It offers opportunities for people to acquire useful skills – not only in building, but also in organizing in the widest sense. People have gained self-confidence from working with others and from dealing with the authorities and professionals.

The barriers to self-build

The principal barriers to the further development of the self-help sector are, I believe, the difficulty of gaining access to land at an affordable price. Land ownership is concentrated in the hands of relatively few people in Britain. This legacy of the role of the church, the state and the nobility which culminated in the agricultural enclosures of the 18th century is in contrast to many other economies which have retained small-scale ownership which originally arose from peasant farming. In the UK, not only is much of the land controlled by the few, but its use is severely limited by planning legislation which prevents residential development in most places, restricting it to existing towns and villages and areas with services.

Also important is the assumption by officialdom that ordinary people cannot be trusted; that they cannot do things for themselves. This is a cultural phenomenon which is not common in places like

North America or northern Europe, where co-operation and individual enterprise are more highly valued. My experience is that people want to do things for themselves properly, and want to make sure that their house is safe and well built – unlike some mass-housing providers.

Authorities with targets to meet within tight budgets prefer to deal with large organizations rather than a multitude of small ones, in the belief that this approach necessarily leads to efficiency – which it often does not. Rather, it tends to entrench the inherent conservatism of the industry, which can make good profits with minimum effort by doing what they have always done. On the other hand, those developers who are looking to the future will be well placed with the necessary experience and expertise to supply the growing demand for low-environmental-impact homes.

It is also true that the majority of people are not aware of the concept of a low-environmental-impact 'green' house, and so the demand and market for low-energy, sustainable homes is not developed. They are probably more aware of organic food and the pollution generated by road traffic. Nevertheless, this situation could change very quickly – the TV series 'Jamie's School Dinners' has changed how we think about what we eat – and the 'Grand Designs' programme could do the same for how we build.

If the market was demanding low-energy sustainable homes, then the developers would build them. Standards in Scandinavia, the Netherlands and Germany are higher than in Britain, and no one there would dream of building a house without incorporating sophisticated energy-saving measures – in much the same way that no one in Britain or elsewhere would dream of building without designing for fire safety. An established market would make sustainable building products more readily available and much cheaper. A modern central heating system with a sophisticated condensing boiler is now far cheaper than heating with open fires, each with its fireplace and chimney, which used to be a far more economical solution.

More and more people are building for themselves, which is good for them and good for society as a whole – but let us look forward to a day when most of those people are building energy-efficient, 'green' homes; homes which are comfortable and affordable to run, cost-effective to build, with a long useful life cared for by the residents, homes with a minimum demand for energy, water and materials. There are many people who would build for themselves if the opportunities were available.

How to specify green

This chapter provides contacts for materials and products that are useful in reducing environmental impacts, but which may not be easy to find using mainstream sources of information and supply.

How to specify green

The materials and products detailed below are intended to cover the principal elements of construction. The selection is based on my personal preferences rather than on quantifiable measures of performance or cost: it is based on what I have found to combine a reasonable balance of robustness, availability, buildability, low environmental impact and value for money. The emphasis is on building construction.

Foundations

Individual pad foundation up to 600mm in diameter and around 1400mm deep can be drilled out with an auger designed for making fence post-holes. A small self-contained machine called a Skidster can be hired from **M W Plant Hire Limited (01279 771803)**.

Structure

Frame structures can reduce the embodied energy; timber is the low-energy material to use for the frame, and recycled or local timber is preferred to imported material. Douglas fir or larch are more durable than other softwoods. Kiln-dried timber is stable, unlike green unseasoned material which is liable to twist, warp and split. Timber production in Britain is concentrated in the highland fringes. Two sawmills that can supply such material are **Charles Ransford & Sons (01588 638331, www.ransfords.co.uk)** and **Pontrilas Timber (01981 240444, www.pontrilastimber.co.uk)**. There are other mills that can supply this type of material, especially in Scotland.

Engineered timber beams can reduce the amount of material used and are produced from waste and low-grade material. I-beams are manufactured in Britain by two companies: **James Jones & Sons Ltd (01981 240444, www.jji-joists.co.uk)** and **Finnforest**

UK (020 8420 0777, www.finnforest.co.uk). Finnforest also supply Laminated Veneer Lumber, (LVL), which is manufactured from thin veneers glued together vertically to make what is in effect a beam made of plywood and is suitable for more heavily loaded members. LVL beams are manufactured in sizes to match I-beam joists.

Masonite I-beams are manufactured in Sweden and imported by **Panel Agency Ltd (01474 872578, www.panelagency.com)**. They have a web of structural grade hardboard made from compressing sawdust under high pressure and temperature. It is environmentally good because it relies on the natural cellulose in the timber to bind it together, rather than manufactured adhesive (as in the other I-beam products which use Oriented Strand Board).

Another form of engineered timber is Eco Joists, which are manufactured using light steel struts to connect the flanges. This permits large diameter ducts, such as are necessary for a whole house ventilation system, to pass through the floor construction without difficulty. They are supplied by a network of fabricators – contact **Gang-Nail Systems Ltd (01252 334691, www.eleco.com/gangnail)** for a local supplier.

OSB structural sheathing for timber construction which is FSC-certified is supplied by **SmartPly Europe Ltd (01322 424900, www.smartply.com)**.

A very useful material for breathing timber-frame construction is fibreboard which is tongued and grooved on all four edges. This enables the boards to be fixed without fully supporting all edges on noggings which saves work and material, and reduces cold bridging. Bitroc is supplied by **Hunton Fiber UK Ltd (01933 682 683, www.hunton.no)**.

Another useful product for timber-frame construction is a breathable membrane that can be used to provide a vapour control and air-tightness layer. A range of suitable products is supplied by **Fillcrete Ltd (01482 635397)**.

Masonry construction can be improved by using lime for mortar and renders. This creates a more vapour-permeable construction which is at less risk of condensation, is less prone to cracking from movement, and permits bricks to be recycled on demolition. Lime products can be obtained from **Lime Technology Ltd (0845 603 1143, www.lime-technology.co.uk)** and **The Lime Centre (01962 713636, www.thelimecentre.co.uk)** in England, **Ty-Mawr Lime (01874 658249, www.lime.org.uk)** in Wales and **The Scottish Lime Centre (01383 872722, www.scotlime.org)** in Scotland.

Insulation

Natural fibre insulation materials generally have very low embodied energy, derive from renewable resources, and are preferable to inorganic materials such as mineral wool which derive from non-renewable but nevertheless plentiful raw materials; these, in turn, are preferable to foam plastic materials, which tend to have relatively high embodied energy and which are derived from petrochemicals. Cork insulation has been used for many years especially in warm-roof construction as it can carry load from maintenance access. Other natural fibre materials include cellulose fibre made from recycled newspaper. This has either to be blown into a cavity, which is not always possible, or used as loose loft insulation. You will need to contact the supplier, **Excel Industries Ltd (01685 845200, www.excelfibre.com)** to put you in touch with a contractor who has the equipment to blow in the material.

Cellulose is also available in the form of batts which can be installed in other situations. The product is more expensive, but it can be installed without special equipment. A small proportion of artificial fibres can be incorporated to render the batts flexible in all directions. Cellulose fibre batts are available from **Construction Resources (020 7450 2211, www.constructionresources.co.uk)** amongst others. They also supply wood-fibre insu-lation boards, which are rigid and can be used in loadbearing applications – in floors, for example. Wood-fibre boards are denser than other similar insulation material, and have the property of increasing the thermal capacity of the construc-tion, which can reduce the risk of lightweight buildings overheating in summer and cooling down too quickly when the heating goes off in winter. Wood-fibre boards and cellulose batts are similar in price; flax batts are another alternative, but are twice the price with no particular advan-tages.

These days, wool is essentially a waste product of meat production. It has been processed for use as insulation in New Zealand and Germany for some time and **Second Nature UK Ltd (01768 486285, www.secondnatureuk.com)** is now producing wool insulation in Britain. It is a very good material to handle, unlike mineral wool which irritates the skin and cellulose fibre which is very dusty. Wool also has the property of being springy so that you can compress it slightly and it will spring back and fully fill the space, reducing the risk of gaps being left in the insulation causing cold spots.

Hemp fibres can be used as loose insulation in roofs and lofts, or for walls and ground floors in combination with lime. The material is supplied by **Hemcore Ltd (01279 504466, www.hemcore.co.uk)**.

Pitched-roof coverings

Wood shingles are a renewable, low-embodied-energy roof covering. Most are made of cedar and are imported; locally produced shingles reduce the embodied energy resulting from transport and stimulate local woodland industries. They can be obtained from **Rawnsley Woodland Products (01208 813490, www.cornishwoodland.co.uk)**. UK-made oak shingles, which are more durable but twice the price, can be supplied by **Low-Impact Living Initiative (01296 714184, www.lowimpact.org)**.

Flat-roof coverings

The preferred materials for flat-roof coverings are either EPDM (Ethylene Propylene Diene Monomer) or TPO (Thermoplastic Polyolefin, also known as Flexible Polypropylene Alloy – FPA). TPO membranes are thermoplastic single-ply roof membranes constructed from ethylene propylene rubber. The roofing differs from EPDM in its welded joints and its improved resistance to chemical and biological attack. EPDM is an elastomeric membrane also made from ethylene propylene rubber. Joints are made either with adhesive or jointing tape. EPDM membranes are supplied by **Firestone Building Products** (01606 552026, www.firestonebpe.com). A number of manufacturers can supply TPO membranes including **DRC Polymer Products** (01582 607718, www.drc-polymers.com), **Flag UK Ltd** (01428 604500, www.flag.it), and **Bailey Roofing Systems** (01444 244330).

Generally, roofing manufacturers will only supply material through a network of approved roofing contractors; they will not supply material for self-build application because they are concerned to protect their reputation from roof failures. DRC Polymer Products may supply material, however. The only company I am aware of that has been willing to supply material for self-help fixing is **Kaliko Roofing Systems** (01604 812040, www.kaltek.co.uk). Unfortunately this is a PVC membrane, but it does allow you to fix the roof membrane yourself using a hot-air gun to weld the joints. This will save money and eliminate any problems with co-ordination with subcontractors, allowing the work to proceed in stages, for example.

Windows

A large choice of high-performance windows and doors is available, made in Scandinavia. They have invested in modern automated factories which can produce made-to-measure windows at a very competitive price. Suppliers include **O Windows UK Ltd** (01603 258890, www.outline.dk), who offer a competitive aluminium-clad window, **Nor Dan,** (Scotland 01698 383364, England and Wales 01452 883131, www.nordan.co.uk).

The Swedish Window Company (01787 467297, www.swedishwindows.com) can supply very good timber sliding patio doors, and **Panel Agency Ltd** (01474 872578, www.panelagency.com) supply Domus windows from Finland that produce a competitive 2+1 double-sash window. **Rationel Windows** (01869 248181, www.rationel.co.uk) are often particularly competitive.

Sashless Window Co Ltd (01609 780202, www.sashless.com) produce a range of windows in Britain similar to high-performance Scandinavian designs. The standard specification is not quite as sophisticated in some respects, you will have to specify factory finish to get full finish to the glazing rebates and triple-locking espagnolettes will have to be specified as an extra. They are sometimes not competitive on price as the factory is not as highly automated.

There is now a very well specified range of windows available made in Britain from FSC-certified timber with Boron timber treatment, warm-edge glazing and organic stain finish. They are competitively priced and available from the **Green Building Store** (01484 854 898, www.greenbuildingstore.co.uk).

Paints, stains and clear finishes

A wide range of microporous paint, stains and clear finishes for timber are supplied by **Osmo** (01296 481220, www.osmouk.com). These products offer a balance between low environmental impact, cost and ease of use. The 'One Coat Only' stain finish is very effective, and as its name suggests can be used in a single coat in many applications, which saves time and money overall, although the initial cost of the material may be higher than other products.

Natural Building Products (01844 338338, www.natural-building.co.uk) supply a range of competitively priced emulsion paints, and **Auro** (01452 772020, www.auro.co.uk) emulsion smells delicious due to the citrus oils it contains.

Keim Mineral Paints Ltd (01746 714543, www.keimpaints.co.uk) supply masonry paint suitable for render and other external and internal surfaces. The paint forms a chemical microcrystalline bond with the substrate, becoming an integral part of the surface. This ensures an extremely long-life protective finish, eliminating the need for regular redecoration, and allows moisture vapour to pass through it whilst preventing the ingress of driven rain. There are buildings in Switzerland and Germany with the original paint still looking good over 100 years later.

Rainwater goods

PVC is to be avoided. Alternatives include aluminium and better (galvanized steel), obtainable from **Lindab** (0121 585 2780, **www.lindab.com**).

Sanitary ware

Green Building Store (01484 854898, www.greenbuildingstore.co.uk) supplies a very water-efficient WC, the Ifo Cera ES4. **Hans Grohe** (0870 770 1972, wwwhansgrohe.co.uk) supplies low-water-use showers and spray taps.

Pipework

Polyethylene pipework is now available as an alternative to PVC for above-ground drainage. It uses similar materials and technology to that used for underground gas and water mains which involves special tools to make the welded joints, so you may need to hire the tools or employ a specialist. **Geberit UK** (01622 717811, **www.geberit.co.uk**).

Polybutylene (PB) is the material used for the pipework of the most commonly available push-fit plumbing system for heating and hot and cold pipework, Hep2O. The environmental impact is substantially less than copper, which is the most common material used for pipework. Stainless steel can be used for any parts of the system that may be exposed internally. Hep2O is manufactured by **Hepworth Plumbing Products** (01709 856400, **www.hep2o.co.uk**), who will supply a list of stockists.

Ventilation

Passive stack ventilation can be obtained from **Aereco Ventilation Ltd** (024 7663 1176, **www.aerecovent.co.uk**).

Floor finishes

Linoleum is the preferred finish for kitchens and bathrooms and is supplied in the UK by **Forbo-Nairn Ltd** (01592 643777, **www.forbo-flooring.co.uk**).

Natural cork has a warm look and feel. It is a renewable natural resource with minimal environmental impacts, however, most cork sheet and tiles are encapsulated with a layer of PVC. Natural cork is perfectly durable if protected with a suitable sealer such as prolix oil by **Osmo UK** (see above). Unsealed tiles are supplied by **Nicoline Ltd** (01772 314665, **www.nicoline.co.uk**).

Alternatively, Osmo UK supply 300x900mm cork panels with T&G joints on a MDF backing presealed with Prolix Oil. This costs more than untreated tiles, but it avoids (a) the need for the ply backing that would be required if tiles are laid on existing floorboards, and (b) the sealing process, which requires two coats of sealer, each of which takes quite a time to dry.

Construction Resources (see above) supply low-environmental-impact natural fibre carpets and adhesive, which avoids the need for high levels of volatile organic compounds (VOCs).

Water butts

Recycled plastic former fruit juice containers are supplied for use as water butts by **Smiths of the Forest of Dean Ltd** (01594 833308, **www.smithsofthedean.co.uk**), amongst others.

Other sources of information

There are a number of sources of information on low-environmental-impact materials and products. Principal amongst them is **Construction Resources** (020 7450 2211, **www.constructionresources.co.uk**).

This is a centre for ecological building, bringing together under one roof in central London a unique range of building products and systems. They sell building materials and systems of a wide range of complexity, from the most basic unfired clay bricks and plasters, to the most sophisticated heating control systems providing maximum comfort with minimum use of non-renewable fossil fuel.

The building includes three floors of display, a seminar room, a warehouse with trade counter and advice counters staffed by specialists in each product area. The centre is mainly directed towards the building trades, architects, engineers and other professionals, but the building is also open to the general public. All the products on display are available for sale from the centre. Both introductory and specialist seminars are open to the general public as well as to professionals, and they have a range of product information sheets available both online and for collection in person.

The National Green Specification (NGS, **www.greenspec.co.uk**) provides free access to very detailed specification clauses incorporating the principles of green construction for all elements of construction, together with comparisons from an environmental point of view of different materials and also product information including manufacturers' contact information.

Building for a Future is a good quarterly magazine devoted to green building, which carries a good many advertisements for green building products as well as articles centred on practical eco-building. It is published by **The Green Building Press** (01559 370798, **www.newbuilder.co.uk**). They also run a web-based, product information database called **GreenPro**, accessible from the same website. Access requires a one-off £10 payment. The information is basic, and no attempt is made to report on the relative merits or disadvantages of particular products, but it does nevertheless provide a source of contacts for green products.

The home page of the National Green Specification website, which gives specification and product information on low-environmental-impact building products.

Useful links, references and contacts

This chapter lists books and websites that can provide information on all aspects of green building, including general principles and specific advice on design and sustainable construction.

Useful links, references and contacts

The basics of self-build are not covered in this book. However, *Building your own Home* by Murray Armor and David Snell (Editor), Ebury Press, is published annually and has good mainstream coverage of self-build. For a very practical approach to basic self-build I suggest *The Housebuilder's Bible: An Insider's Guide to the Construction Jungle* by Mark Brinkley and published by Rodelia Ltd, Cambridge.

There are a number of magazines aimed at self-builders. The principal ones are *Build-It, Self-Build and Design* and *Homebuilding & Renovation*. They offer case studies, advice and information on finding building plots.

The *Ecology Building Society* (0845 674 5566, www.ecology.co.uk) lends to sustainable building projects and has funded many eco-self-build projects.

For a wide context within which to set self-help housing, I would recommend **John Turner**, who argues that housing can only be successful if residents have a role in the housing process in *Housing by People; Towards autonomy in building environments*, Marion Boyars. **Colin Ward and Dennis Hardy** described the history of the plotland developments between the wars which were largest manifestation of self-build in Britain in their book *Arcadia for All; the Legacy of a Makeshift Landscape*, Five Leaves Publications. **Simon Fairlie** outlines measures to create sustainable communities in rural areas in his book *Low Impact Development; Planning and People in a Sustainable Countryside*, Jon Carpenter Publishing.

The Building Regulations together with the detailed Approved Documents which explain how to comply with the regulations can be downloaded free from **www.planningportal.gov.uk**.

Chapter 1. Why self-build, and why build green?

How We Can Save the Planet by Mayer Hillman, Penguin Books, explains the issues of climate change very clearly, provides a context for sustainable housing and argues that carbon-rationing is indeed the way to save the planet. *The Rough Guide to Sustainability* by Brian Edwards published by RIBA Enterprises pulls together all the disparate, complex strands of sustainability into one simple reference source. It sets out the context underlying sustainable principles, as well as outlining the approach that designers must adopt to meet 21st-century expectations of responsible architecture.

The UK Climate Impacts Programme (UKCIP) provides scenarios that show how our climate might change and co-ordinates research on dealing with our future climate. For details of climate change impacts on buildings see *Workshop report: Climate change and the built environment research fora*, which is available for downloading from **www.ukcip.org.uk/built_enviro/built_enviro.html**.

Essential references for 'green' house-building include *The Whole House Book* by Pat Borer and Cindy Harris published by The Centre for Alternative Technology, which gives very good coverage of sustainable building techniques. They also wrote *Out of the Woods; Ecological Designs for Timber-Frame Housing*, also published by CAT, which details how to design low-impact timber homes. *Ecohouse 2, A Design Guide* by Sue Roaf, Architectural Press includes detailed case studies from around the world.

The Handbook of Sustainable Building by Anink, Boonstra and Mak (James & James London, 1996) describes the Environmental Preference method for making decisions on the selection of building materials. It is based on practice in the Netherlands but most of the contents apply in Britain. *The Green Building Handbook*, Volumes 1 and 2 by Woolley *et al* (E & FN Spon, London,

1997 and 2000) assesses the environmental impacts of different materials and products and is a good companion to the *Handbook of Sustainable Building*. The *Green Guide to Housing Specification*, **Building Research Establishment**, Watford, gives an environmental grading of a number of common forms of construction for the various elements of the building.

Building for a Future is a good quarterly magazine devoted to green building which carries a good many advertisements for green building products as well as articles centred on practical ecobuilding. The website **www.buildingforafuture.co.uk** has free downloads of back issues. It is published by **The Green Building Press** (01 559 370 798, **www.newbuilder.co.uk**). The Green Building Press also publishes *The Green Building Bible* which is a good source of information and in particular includes the membership directory of the **Association of Environmentally Conscious Builders** (**www.aecb.net**), which promotes sustainable building and has a membership which includes builders, architects, manufacturers, suppliers as well as builders and contractors.

Chapter 2. Who has successfully built their own green self-build house?

The Hockerton Housing Project (01636 816902, **www.hockertonhousingproject.co.uk**) publish information on the Hockerton project and other sustainable housing projects in Britain. They also run guided tours and a programme of seminars.

An Information Guide to Straw Bale Building for Self-Builders and the Construction Industry is published by **Amazon Nails** (01706 814696 **www.strawbalefutures.org.uk**), who are straw-bale builders who also run courses and offer consultancy in straw-bale building.

Report on the Construction of the Hemp Houses at Haverhill prepared by the **Building Research Establishment** is a detailed analysis of the performance of this form of construction, and can be downloaded from **www.bre.co.uk/pdf/hemphomes.pdf**.

Chapter 3. Designing a good house

I commend the philosophy of building outlined in **Christopher Alexander**'s book *The Timeless Way of Building*, New York, Oxford University Press, 1977. His *Pattern Language: Towns, Buildings, Construction*, New York, Oxford University Press, 1977, is an invaluable sourcebook when considering the design of a house from the large scale to the smallest details.

Chapter 6. Building for longevity

How Buildings Learn: what happens after they are built by **Stewart Brand** (Viking, 1994) is an interesting investigation into how the long-term use of buildings could influence their design.

For details of the **Lifetime Homes** standard, which was developed by the Joseph Rowntree Foundation, see **www.jrf.org.uk/housingtrust/lifetimehomes**.

Chapter 7. Reducing energy in use

The Energy Efficiency Best Practice Programme publishes a comprehensive range of guides accessible at **www.est.org.uk/housingbuildings/publications** on new and refurbished housing, and case studies which can be downloaded free.

The National Energy Foundation provides information, practical advice and information on suppliers of renewable energy products and services. **www.greenenergy.org.uk**.

The British Wind Energy Association (BWEA) is the trade association of the UK wind industry: **www.bwea.com**.

The Solar Trade Association represents the solar hot water industry: **www.solartradeassociation.org.uk**.

Good Energy at **www.good-energy.co.uk** supplies electricity from 100% renewable sources.

Future Forests is an organization set up in

Britain to promote and carry out carbon offset in an environmentally responsible manner: see **www.futureforests.com**.

Chapter 8. Reducing environmental impacts

The sustainable use of timber

The Central Point of Expertise on Timber Procurement (CPET) website at **www.proforest.net/cpet** provides detailed information and advice on how to meet the UK Government's timber procurement policy. It assesses the evidence of legal and sustainable forestry practice offered by the principal certification schemes – **Forestry Stewardship Council (FSC), Programme for the Endorsement of Forest Certification (PEFC), Sustainable Forestry Initiative (SFI)**, which is applicable to timber from the USA and Canada, **Canadian Standards Association (CSA)** and **Malaysian Timber Certification Council (MTCC)**.

The Good Wood Guide by **Friends of the Earth** for general information and sustainable timber products from **www.foe.org.uk**.

Reuse and recyclability of materials

Eco-renovation by **Edward Harland** (Green Books, Totnes, revised edition 1999) is a step-by-step guide to allow the 'green' homeowner to make a small but significant contribution to the well-being of the global environment.

www.wastewatch.org.uk for background on different recycled products and lists of suppliers,

www.nrf.org.uk (National Recycling Forum) for a product guide with lists of suppliers,

www.recycle-it.org for information on recycling timber,

www.bre.co.uk/waste takes you to the **Materials Information Exchange**, which lists needs and offers of reclaimed building materials (however this site has relatively little posted at the time of writing),

www.salvo.co.uk takes you to a site with details of dealers in architectural salvage.

Chapter 9. Reducing harmful impacts on health

Daylighting

Desktop Guide to Daylighting by **Building Research Establishment**, Watford.

Noise

Quiet Homes, A Guide to Good Practice by **John Seller** (BRE, Watford, 1998).

Toxicity

The London Hazards Centre has comprehensive information on pollutants and the health impacts of poor air quality which may arise from inadequate ventilation at **www.lhc.org.uk**.

Electro-magnetic fields

The charitable organization **Powerwatch** has information on EMFs on **www.powerwatch.org.uk**.

Chapter 10. Reducing waste

Construction waste

Building Research Establishment, www.bre.co.uk, for site waste audit system **SMARTwaste** and **The Materials Information Exchange** with requests and offers of reclaimed materials.

www.nrf.org.uk National Recycling Forum for a product guide and list of suppliers.

www.salvo.co.uk is a site with details of dealers in architectural salvage.

Domestic waste

www.letsrecycle.com has information on recycling in general.

Chapter 11. Reducing water consumption

Water conservation

Conserving Water in Buildings Fact Cards, The Economics of Water-Efficient Products in the Household and *A Study of Domestic Greywater Recycling*, all published by the **Environment Agency** www.environment-agency.gov.uk.

Sewage Solutions by **Grant, Moodie and Weedon, Centre for Alternative Technology**, www.cat.org.uk.

Low-Water Gardening; Creating and running the ideal garden with less water by Lucas, J. M. Dent, London, 1993, and *The Dry Garden*, Chatto, Orion, 1998.

Storm Water

Sustainable Urban Drainage Systems: an introduction free from **The Environment Agency** and **Scottish Environment Protection Agency**. Good introduction to the principles and a pointer to further information.

Sustainable Urban Drainage Systems Manuals on Best Practice and Design from **The Construction Industry Research and Information Association (CIRIA)**, www.ciria.org.uk.

An assessment of flood risk for a particular area is available on the **Environment Agency website** at www.environment-agency.gov.uk.

Sewerage treatment

Sewage Solutions by **Nick Grant, Mark Moodie, Chris Weedon, Centre for Alternative Technology**, Machynlleth, 2000. This is a very readable and practical description of the theory and practice of on-site sewage treatment.

Good Building Guide GB42. Part 1. Reedbeds: application and specification. Part 2. Reedbeds: design, construction and maintenance by **N. Grant and J. Griggs**, BRE, Watford, 2000, available from CRC Ltd, 020 7505 6622.

Lifting the Lid by **P. Harper and L. Halestrap**, Centre for Alternative Technology, Machynlleth, 2000. The definitive UK dry toilet book; also covers household nutrient recovery and composting.

Chapter 13. Designing a sustainable garden

How to make a Wildlife Garden by **Chris Baines** (London, 1985).

www.wildlifetrust.org.uk gives details of local wildlife trust offices. **The Wildlife Trusts** can give advice on retaining and enhancing wildlife and biodiversity.

Index